Engineering Pedagogy Towards Outcome-Based Education

With the growing environment and consciousness of "outcome-based education," the importance of this subject has increased manyfold. Unfortunately, there is little information on engineering pedagogy available outside of scattered journal articles, conference and symposium proceedings, workshop notes, and government and company reports. This book overcomes these difficulties by presenting, in a single volume, many of the recent advances in the field of engineering pedagogy and its recent developments.

Engineering Pedagogy Towards Outcome-Based Education provides a systematic approach to explicit fundamentals as well as recent advances in the area. It incorporates various case studies for major topics as well as numerous academic examples. Each chapter contains many state-of-the-art techniques required for practical engineering applications.

This book serves as a useful source of information for practicing academicians and specialists as well as academic institutions working on the subject.

H0234948

Engineering Pedagogy Towards Outcome-Based Education

Edited by
Kaushik Kumar

CRC Press is an imprint of the
Taylor & Francis Group, an **informa** business

First edition published 2023
by CRC Press
6000 Broken Sound Parkway NW, Suite 300, Boca Raton, FL 33487-2742

and by CRC Press
4 Park Square, Milton Park, Abingdon, Oxon, OX14 4RN

CRC Press is an imprint of Taylor & Francis Group, LLC

© 2023 selection and editorial matter, Kaushik Kumar; individual chapters, the contributors

ISBN: 978-0-367-53743-2 (hbk)
ISBN: 978-0-367-53745-6 (pbk)
ISBN: 978-1-003-08316-0 (ebk)

DOI: 10.1201/9781003083160

Typeset in Times
by KnowledgeWorks Global Ltd.

Contents

SECTION I   State of Art

SECTION II   Ethics

SECTION III   Tools and Methodology

SECTION IV Applications

Preface

The editor is pleased to present the book *Engineering Pedagogy Towards Outcome-Based Education*. The book title was chosen in view of the current importance of the subject. This book will be a key to support the Washington Accord, which specifically focuses on academic programs that deal with the practice of engineering at the professional level. The Accord acknowledges that accreditation of engineering academic programs is a key foundation for the practice of engineering at the professional level in each of the countries or territories covered by the Accord.

Pedagogy is defined as "the study of the methods and activities of teaching". Engineering pedagogy, as the title says, is closely linked to the technical and pedagogical sciences. It is also associated with the preparation of teachers who teach in engineering schools and universities. A central role of engineering didactics is primarily dictated by the concept of technology, or in other words, the primary function of the technology is to transform the natural world. In past 100 years the concept of engineering education has moved from hands-on/practically oriented education to education integrating information and computational and communications technology.

Object-related reasoning concerning the design of learning and teaching processes in academic engineering education illustrates the scientific character of related questions. Verifiable pedagogical and/or psychological qualifications are legally fixed requirements for a teaching career at all levels and in all types of schools of general and vocational education. In contrast, in the sector of higher education, it is assumed that lecturers have teaching abilities due to their high academic qualifications, but evaluation results partially counter this assumption. A major reason for this is the complexity of the influence factors and relationships concerning the design of a demand-oriented education in engineering sciences.

Each chapter has been written by recognized experts and academicians practicing and following the outlook of the concept, and hence the book contains many state-of-the-art techniques required for practical applications of engineering pedagogical and didactic aspects. Thus, this book should serve as a useful source of information for practicing academicians and specialists, as well as for academic institutions working on the subject. At the core of the book are several application areas where pedagogy concept can be applied, each treated as a complete chapter. This book also provides an introduction to the approach and mindset that give the readers a taste of the techniques and methods of the same. The application chapters provide discussion and reflection that will lead the reader into a deeper understanding of the nature of outcome-based education. I am confident that this book will serve as an easy-to-understand guide to facilitate anyone to learn and apply the methods and tools to generate innovative ideas and grasp the low-hanging fruit in the short term and design a better education system in the long run for the betterment of society as a whole.

The chapters of the book are segregated in **four Sections** namely **Section I: State of Art, Section II: Ethics, Section III: Tools and Methodology,** and **Section IV:**

Applications. Section I contains **Chapter 1** to **Chapter 3**, whereas **Section II** comprises **Chapter 4** only; **Section III** consists of **Chapter 5 to Chapter 7**, and **Section IV** has **Chapter 8** to **Chapter 11**.

Section I starts with **Chapter 1** providing the state of art of pedagogy towards outcome-based engineering education. Engineering education is now confronting various challenges that are the barriers in the progress of educational establishment as well as nation-building. Consequently, the demand and desire to adopt the new teaching and learning methodology have become the necessities in today's technological era. In this context, outcome-based education (OBE) is the newly adopted philosophy in education. This chapter is a modest attempt to enlighten readers on the recent development and newly adopted methodologies towards engineering pedagogy. A COVID-19 case study has also been introduced to analyze how the massive pandemic has affected the education sector around the globe and with the usage of information and communication technologies (ICT) how this world has managed to continue the teaching and learning process (TLP) along with the challenges faced in the sudden transformation of classroom leaning to online digital learning.

Chapter 2 talks about lifelong learning practices. Learning is based on change, and change is what propels it ahead. Both are intrinsically related and go together. Change creates gaps between what is and what will be, or between what happened and what is happening now. Learning is the ability to influence the future by allowing a person to make decisions they would not have been able to make otherwise. In the modern digital economy, the need of lifelong learning for active citizens cannot be overstressed. The development of generic competencies, as well as specific competencies in areas like engineering, should be incorporated into all courses. New technologies are currently influencing instructional design and delivery of educational program. The chapter examines learning in young and old individuals around the world, at work and outside of work, and offers advice on how to think about aging, personal development, overcoming barriers, and innovation in the current era of outcome-based education.

The next chapter, **Chapter 3**, also the last chapter of the section, features transformational leadership, the new mantra for pedagogy in higher education. The chapter outlines basic requisites to be a transformational leader, along with details in evolution and change in the transformation. The strategies to improve leadership qualities are discussed under the purview of teaching, the learning process, and research excellence. The highlight of the work is on the Cambridge Model that provides all the required insights into the student outcomes. The chapter concludes with an annotation to the Industry 4.0 constituents and the UN sustainability goals that every transformational leader must be aware of in order to provide global and national ranking and accreditation to an institute or organization.

The best practices of engineering are the outcome of ethical principles of engineering ethics. It is an obligation set by the system for engineers towards society and the profession. There is an impact of services provided by engineers on the standard of life for all people. They need to practice honesty, impartiality, fairness, and equity. Their service requires dedication towards the protection of the general public health, safety, and welfare of the full society, and **Chapter 4**, the only chapter of Section II, provides a systematic approach towards an ethical perspective on engineering education. Earlier engineering education was focused on technical specifications excessively, and now the time has come to be creative and develop curriculum that matches the frequent changes and demands of technology, abolishing the gap between creativity, innovation, and engineering. Hence engineering curriculum should have the combination of these parameters. The moral perspective of education concerns engineering design ethics of technological products and its systems and sustainability. The curricula must mention all these concerns in engineering education, and hence engineering education must practice a versatile approach rather than being just applying the standard stereotype.

In **Chapter 5**, the progress in adoption of technology in engineering education in developing countries is presented. The chapter begins by providing a general outlook of integration of technology in engineering pedagogy in developing countries, the potential it offers, and its success in such areas. A relationship between Bloom's taxonomy and technology in engineering teaching is also illustrated. Using a case study of the authors who were involved in online teaching at Dedan Kimathi University of technology (DeKUT), a university in Kenya (a developing country), lessons in online teaching, which developing countries can adopt to ensure continuity in learning even during times of the pandemic, are presented. The chapter concludes by providing the implications of the study to the university teaching and training with the advancement in technology and ever-increasing uncertainties in the modern world.

The next chapter, **Chapter 6**, highlights the significance of outcome-based assessment and evaluation-centered performance indicators for continuous quality improvement of engineering education. The chapter quotes the ABET and other global standards to design the methodology for assessment and evaluation (A&E) of student outcomes (SOs), which forms the basis of the continuous improvement activities. Moreover, the curricula in engineering programs have to be aligned with the outcomes determined by the respective criteria by applying, for example, performance indicator (PI)-breakdown–based assessment, facilitating attainment of more realistic, meaningful, and easily assessable results in their A&E studies for providing a genuine picture of the relevance, accuracy, and utility of the selected assessment tools. In this chapter, some insight has been provided in an environmental engineering undergraduate program.

The last chapter of the section, **Chapter 7** depicts the current status of low employability of fresh engineering graduates, as well as the absence of premier engineering institutions of India in global rankings. It is becoming increasingly clear that improving the quality of engineering education in India will be impossible until "quality" is mentioned in terms of "learning outcomes," apart from grades, and judged consequently. The authors predict that choosing outcomes-based education will benefit all stakeholders – scholars, parents, engineering colleges, employers, and government – not just in terms of improving the employability of graduating engineers but also in terms of improving the overall quality of engineering education and the society as well.

The next and last section of the book, Section IV handles the application part, and the first chapter of this section, **Chapter 8**, talks about information transfer in education, with an example based on teaching lowcost control unit programming.

In this chapter authors have dealt with efficiency of different ways of information transfer, in particular information presenting subject matter of teaching lowcost control unit programming, which is one of the branches on which the further development of each Industry 4.0 society is based. Within their case study, the authors compared efficiency of the information transfer carried out by means of texts, pictures, and video recordings. The research sample of the case study consisted of 24 secondary vocational school students divided into three groups, each of which had the information – subject matter related to lowcost control unit programming – presented in one of the stated ways. Within the case study, in relation to the transfer of the technical information linked with the control unit programming, on the one hand the commonly accepted claim that video is the strongest stimulant for learning and remembering information was not proved, and on the other hand the significance of text in transfer of the corresponding information was declared.

Chapter 9 deals with home assignments and assessments in online learning. The ongoing pandemic caused by outbreak of COVID-19 has brought a paradigm shift in education. Institutes cancelled face-to-face classes and shifted to online. This shift has brought new challenges for students and teachers as well. One such challenge is conducting and assessing assignments and examinations. Proper assessment is needed for assignments and examinations, as the performance of a student in these reflects their progress and understanding of the subject. It is crucial to maintain the standards of education and ensure that the expected outcomes are achieved. So, the current study aims to investigate this problem of assessment of home assignments through a questionnaire floated to undergraduate students of an engineering college in a developing country. For the study, two questionnaires were prepared, one regarding assignments and their assessment and another regarding examinations and their assessment. The responses of students were taken and analyzed. After analyzing the responses, an attempt was made to provide some input for a more efficient process.

The penultimate chapter of the book, **Chapter 10**, aims to comparatively display students' perception and instructors' views in curriculum mapping for attainment of student outcomes (SOs), explore reasons for mismatches, and offer recommendations for improvement. A sample student portfolio, including a student-derived curriculum map, prepared by a voluntary group of senior-year engineering students was utilized and compared to that prepared by the instructors. Comparison of the curriculum map prepared by the students to that constructed by the instructors revealed considerable matching aspects. Students' projection of relating more courses to the SOs were interpreted as a reflection of the broader student perception. Lack of scaling of level of contribution of courses in students' map limited implementation in assessing SO achievement levels. Those pointed to the need for better guidance of students in preparing their academic portfolios so that outputs might be utilized in SO assessment and evaluation properly. As this chapter focuses on students' perception in curriculum mapping for attainment of SOs and explores reasons for mismatches with that of instructors, it calls for specific attention to be paid to the recently implemented student advising courses as effective starting points and useful platforms for the purpose of achieving better guidance of students in preparing their academic portfolios.

The last chapter of the section and the book, **Chapter 11**, describes learning management systems (LMS) as an intrinsic web-based component to support and promote teaching and learning. With the advent of online teaching, the use of information and communication technology (ICT) has opened several avenues for teachers and learners to have a proper LMS, be it paid, or open source, or developed at the institute. And the features and tools embedded in the LMS have made it suitable for online learning and teaching. LMS has also been used as a supplement to face-to face teaching and learning. In this chapter two open-source LMSs, namely EDMODO and Google Classroom, are compared for teaching mathematics and English language in an engineering institution.

First and foremost, I would like to thank God. It was your blessing that provided me the strength to believe in passion and hard work and to pursue dreams. I thank my family for having the patience with me for taking yet another challenge that decreases the amount of time I could spend with them. They were my inspiration and motivation. I would like to thank my parents and grandparents for allowing me to follow my ambitions. I would like to thank all of my colleagues, friends, and associates in different part of the world for sharing ideas in shaping my thoughts. My efforts will come to a level of satisfaction if the students, researchers, and professionals concerned with all the fields related to engineering pedagogy are benefitted.

I, from the bottom of my heart, owe a huge thanks to each and every contributing author, reviewers, editorial advisory board members, book development editor, and the team of CRC Press for their availability for work on this huge project. All of their efforts were instrumental in compiling this book, and without their constant and consistent guidance, support, and cooperation I wouldn't have reached this milestone. Especially during this global pandemic, when all supports were withdrawn, I was elated to find the whole team of CRC Press by my side in support. I salute their dedication.

Last, but definitely not least, I would like to thank Ranjan Kumar, my scholar, and all individuals who have taken time out and helped me during the process of writing this book. Without their support and encouragement I would have probably given up the project.

Kaushik Kumar

Editor

Kaushik Kumar, BTech (Mechanical Engineering, REC (Now NIT), Warangal), MBA (Marketing, IGNOU) and PhD (Engineering, Jadavpur University), is an Associate Professor in the Department of Mechanical Engineering, Birla Institute of Technology, Mesra, Ranchi, India. He has 14 years of teaching and research experience and over 11 years of industrial experience in a manufacturing unit of global repute. His areas of teaching and research interest are quality management systems, optimization, non-conventional machining, CAD/CAM, rapid prototyping and composites. He has 9 patents, 30 books, 27 edited book volumes, 46 book chapters, and 200+ journal and conference publications to his credit (more than 100+ articles in journals indexed in Web of Science collection/h-index 8+/200+ citations, SCOPUS/h-index 8+/170+ citations, Google Scholar/h-index 17+/1300+). He is Editor in Chief, Guest Editor, Editor, Series Editor, and Scientific Advisor for many international journals and conferences. He is also an editorial board member and on the review panels of international and national journals of repute. He has been honored with many awards and honors.

Contributors

Anju Bharti
Maharaja Agrasen Institute of Technology
Delhi, India

Ananya Chepuri
Geethanjali College of Engineering and
  Technology
Hyderabad, Telangana, India

Ebru Dulekgurgen
Istanbul Technical University
Istanbul, Turkey

Alena Hašková
Constantine the Philosopher University
Nitra-Chrenová, Slovakia

Burcak Kaynak
Istanbul Technical University
Istanbul, Turkey

Aryan Kodte
Geethanjali College of Engineering and
  Technology
Hyderabad, Telangana, India

Padmanabhan Krishnan
Vellore Institute of Technology
Vellore, Tamil Nadu, India

A. Kishore Kumar
Sri Ramakrishna Engineering College
Coimbatore, Tamil Nadu, India

A. Udhaya Kumar
Hindusthan College of Engineering and
  Technology
Coimbatore, Tamil Nadu, India

Kaushik Kumar
Birla Institute of Technology
Mesra, Ranchi, India

Ranjan Kumar
Birla Institute of Technology
Mesra, Ranchi, India

Peter Kuna
Constantine the Philosopher University
Nitra-Chrenová, Slovakia

S. Lalitha
Geethanjali College of Engineering and
  Technology
Hyderabad, Telangana, India

F. M. Mwema
Dedan Kimathi University of
  Technology
Dedan Kimathi, Nyeri, Kenya

B. Nagamani
Geethanjali College of Engineering and
  Technology
Hyderabad, Telangana, India

T. Nivethitha
Hindusthan College of Engineering and
  Technology
Coimbatore, Tamil Nadu, India

Miloš Palaj
Secondary Vocational School of
  Polytechnics
Zlaté Moravce, Slovakia

P. K. Poonguzhali
Hindusthan College of Engineering and
  Technology
Coimbatore, Tamil Nadu, India

Bhartipudi Saketh Ram
Geethanjali College of Engineering and
  Technology
Hyderabad, Telangana, India

Shivangi Sahay
Indira Gandhi Delhi Technical University
    for Women
New Delhi, India

N. Subadra
Geethanjali College of Engineering and
    Technology
Hyderabad, Telangana, India

J. K. Tanui
Dedan Kimathi University of Technology
Dedan Kimathi, Nyeri, Kenya

Didem Okutman Tas
Istanbul Technical University
Istanbul, Turkey

Cigdem Yangin-Gomec
Istanbul Technical University
Istanbul, Turkey

Gulsum E. Zengin
Istanbul Technical University
Istanbul, Turkey

Section I

State of Art

Section I

1 Pedagogy Toward Outcome-Based Engineering Education
The State of Art

Ranjan Kumar and Kaushik Kumar
Birla Institute of Technology
Mesra, Ranchi, India

CONTENTS

1.1 INTRODUCTION

Engineering pedagogy has a long historical background toward outcome-based education. The discussion on pedagogical approach in the domain of teaching and research-based education system can be traced to 1851 during the establishment of the Institute for Engineering Pedagogy by Hans Lohmann [1]. The institute was focused on research-based educational approaches toward the design of technological enhancements in teaching and learning. According to Lohmann, the function of "technology is to transform the natural world". The paramount task of the engineers is to develop such technology that could transform the natural world into a

DOI: 10.1201/9781003083160-2

technological world where the expertise of engineers could resolve the technical design problems [2]. Therefore, engineers are essential to be highly qualified in all respects. Natural scientists, on the other hand, spend their time discovering relationships in the world and, as a result, addressing scientific knowledge challenges and invention. Hence, it is notable that invention and discovery demand different approaches to "thinking and different academic training".

Outcome-based education (OBE) has been used in the domain of education at various levels around the world. Since 1990, Australia and South Africa have implemented such OBE programs. Since 1994, the United States has also had an OBE program. In recent years, OBE has become a widely acknowledged ideology in academia, particularly in the subject of engineering. The OBE approach differs from the usual way of evaluating pupils solely on the basis of grades and/or ranks [3].

Further, in 1990, another perspective of engineering pedagogy was developed that was based on the previous understanding of engineering pedagogy and focused on the engineer's ability to be involved in the "social communicative processes toward the modern structure of production and services". Hence, engineers focus on the practical approach while scientists focus on cognitive knowledge [4, 5]. In order to achieve the goal, both engineers and scientists should essentially possess a good imagination that allows them to envision, create and contribute toward the pedagogy of social, cultural, and educational benefits [6]. The word *pedagogy* has been derived etymologically from the Greek word which includes the word *agein,* which means "to lead", as well as the word *pais,* which means "boy" [1]. Thus, the engineering pedagogy refers to focusing on the design and development of teaching concepts and methodologies for future engineers to nourish and cherish their leadership quality toward their role in "changing the structure of production and services". Therefore, pedagogy can be identified as an act of teaching [7] that helps in converting the teacher's knowledge into actions for a better understanding of their students. Hence, students are identified as agents in pedagogy theories, and the instructor is identified as a facilitator. Over the past two decades, scientific and engineering pedagogy practices have advanced dramatically [7–9]. The advancement that took place over the past few years can be estimated on the basis of the graphs delineated in Figures 1.1 and 1.2. Theoretically, as well as practically, both the aspects are essential for a modern engineering course, but the more practical hands-on skill development of a teacher helps in improving the learning ability of students. "Object-related reasoning oriented" teaching and learning techniques are quite helpful in engineering academics. Hence, the exposure of today's engineering academics to the industrial updates and practical approach as per industrial requirements is very important and should be incorporated in engineering curricula and must be a part of institutional training [10]. In 1972, the International Society for Engineering Pedagogy (IGIP) was formed and several universities are currently attempting to develop unique modules in various domains such as engineering education in theory and practice, psychology and sociology, and scientific writing challenges, among others. These gradually led to the creation of a curriculum for an advanced training course in engineering pedagogy. Another challenge in pedagogy research is communication between learners and specialists [7, 11].

In today's technological era of computational and internet technology, the development of course modules via several webinars and online training programs has helped teachers in many ways. But online education and full-time internet-based teaching are unaffordable and impossible in many places. "The demand-oriented advanced training

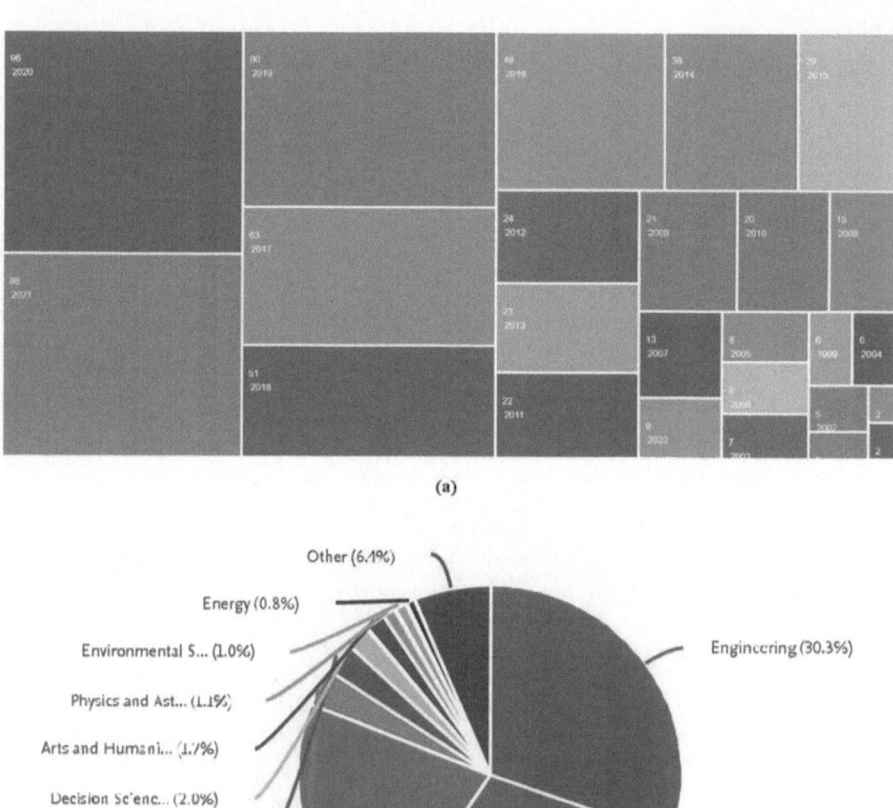

(a)

(b)

FIGURE 1.1 Various documents published in the domain of engineering pedagogy: (a) year-wise publication; (b) subject-area wise publication [12, 13].

is accredited both by the International Society for Engineering Education (IGIP) and by the Scientific Society for Engineering Education (IPW). The IGIP curriculum follows a classical study structure. First, discipline-oriented foundations (pedagogy, psychology, sociology) are taught, followed by selected applications (project work, communication, etc.). The IPW curriculum also focuses on the target group of teachers of engineering sciences. It is a very open curriculum, which can be adapted to the needs of the respective university and the respective engineering sciences". Currently, digital manufacturing and design are providing a critical thought mindset to students and helping them to achieve cognitive exposure toward product-life cycle management. Computerbased design education includes parametric and feature-based modeling techniques, and core concepts of surface and solid modeling are essential for teaching computeraided design. Motion simulations and finite element tools are also used in CAD courses

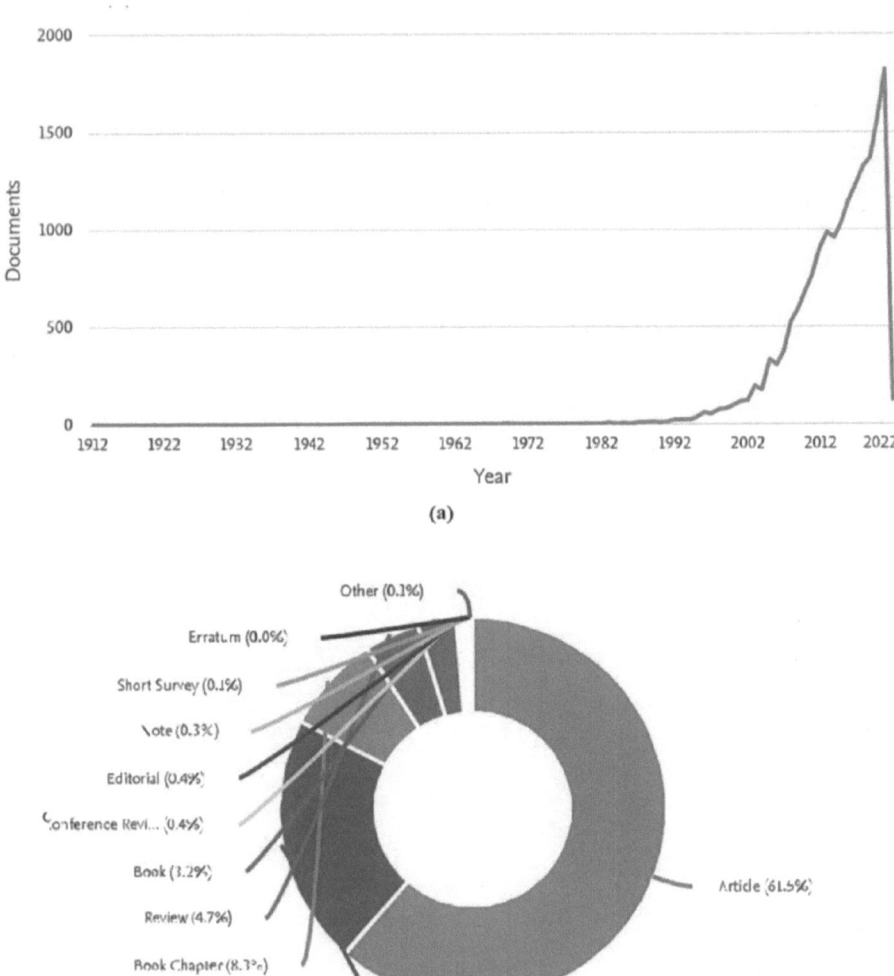

FIGURE 1.2 Various documents published in the domain of teaching and learning-based pedagogy: (a) year-wise publication; (b) documents by type [12, 13].

to attract students' attention and help them apply their learnings in an effective manner [10]. In describing the pedagogy design, the work done by Royalty [14] is appreciated. The instructional design practice allows the students to think critically and creatively to solve human problems. This helps in playing an imperative role in developing a robust educational environment. Similarly, smart manufacturing and automated process planning technologies can be implemented in learning environments to help students comprehend manufacturing processes more efficiently. Content, pedagogy, and technology constitute a knowledge framework, according to a recent discovery.

In pedagogy theory, there exists a difference between content knowledge (CK) and pedagogical knowledge (PK). CK is all about the teachers' understanding and knowledge of the subject matter in which he/she must be a subject matter expert (SME), whereas the PK is all about expertise in delivery of the subject matter or the methodologies and processes adopted to deliver one's knowledge [7]. The overall construction of this current work is as follows: in the very beginning a brief introduction on pedagogical approach in TLP has been provided. In the next section, design-based pedagogy to solve human-centered problems has been summarized, which involves the instructional structures that allow students to address problems through design methodologies. In the next section, various pedagogical tools for quality enhancements have been summarized along with the incorporation of technology in the teaching–learning approach. In the next section, the challenges associated with implementation of such pedagogy tools have been discussed, and for better understanding of such technology-based pedagogy, a case study of COVID-19 has been briefly discussed. Data and graphs related to various surveys have been delineated in order to understand and estimate the importance and requirements of engineering pedagogy in today's technological era. Finally, the conclusion and future scopes have been summarized to design a better quality and outcome-based pedagogy.

1.2 DESIGN-BASED PEDAGOGY

In engineering practice, design is technically a central activity of all engineering procedures [15] and helps in providing quality education to produce good future engineers [16–19]. Simon [20] proposed a definition of design as "a course of action aimed at changing existing situations into preferred ones". Design is all about producing a positive development on which the whole scope of students' activities depends. One of the primary objectives of design pedagogy is to provide a learning-based platform to non-designers for enhancing their creative and cognitive thoughts. Design pedagogy and design thinking are very similar in various ways [21, 22]. Many of the design procedures are "human-centered design". Human-centered design is at the heart of the strategic design approach. Design is practiced by all employees at the company, regardless of their professional experience. Outside of academics, the growth of design has had an impact in the classroom. Many practitioners have returned to higher education, equipped with real-world experience, and paired with teachers who have their own design knowledge to teach design to non-designers [14].

Design process (DP) helps in educating a particular content to non-designers. In doing so, the two main approaches, learning by design (LBD) [23] and design-based learning (DBL) [18], play a vital role. These two approaches have proven to be most effective in using design to support the process of student learning and teachers' instruction. According to the research, teaching design to non-designers is a unique and effective method of education. However, the concept of pedagogy is demonstrated in various other domains. The majority of the work on design as a content conveyer takes place in primary and secondary education. Although that is important, a greater focus on higher education is required [14]. To dig deeper into this design pedagogy, we must first define what it means to teach design to non-designers. The definition must take into account how design is taught in the domain of higher education. Design-based pedagogy (DBP) is a learning field that involves instructional structures that allow students to address problems through design. It has five primary characteristics:

- Target audience—it is directed mostly to non-designers.
- Difficulties—the projects are open-ended and take place in a setting outside of the classroom.
- Multidisciplinary teams—students work in interdisciplinary teams the majority of the time.
- Practice—the way students tackle problems is influenced by a process or set of principles used by designers.
- Uniqueness—improving student creativity is a primary goal.

DBP is a brand new type of education. It resembles previous paradigms but has its own characteristics. There has been no previous research into how instructors create these learning environments. Given its growing popularity, it's vital to know how effective a pedagogy DBP is. Are there any obvious flaws? Is it going to introduce a new set of best practices? These are some DBP-based questions that must be answered. To answer such DBP-related questions, it is necessary to scrutinize the DBP under three developed frameworks, which possess three distinct perspectives:

- instructional design framework (IDF) [24]
- constructivist learning environment (CLE) [25]
- learning space framework (LSF) [26]

All these above frameworks, as shown in Figure 1.3, together endeavor to develop "the overall learning environment" [14]. All these frameworks as well as DBP occur simultaneously.

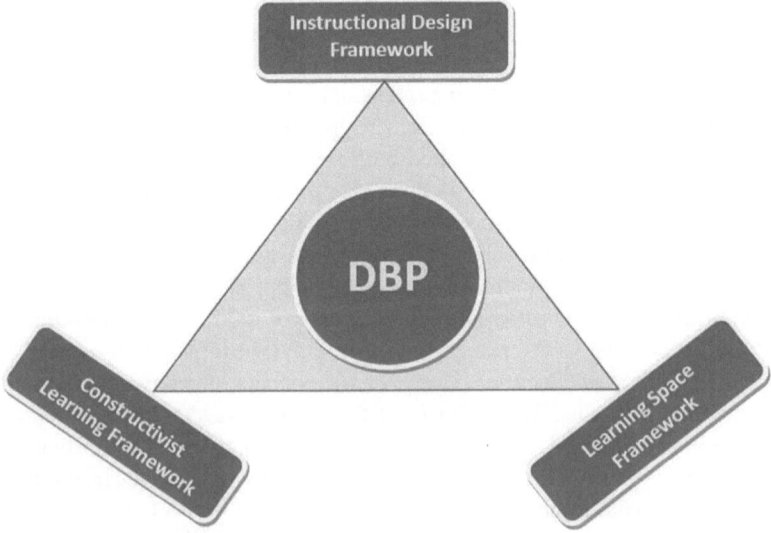

FIGURE 1.3 Schematic of three DBP frameworks.

1.3 PEDAGOGICAL TOOLS FOR QUALITY ENHANCEMENT

In the present technological world, conceptual learning is more likely encouraging students toward critical thinking and directing their interest toward design and development. Service-learning process (SLP) is another aspect of learning pedagogy, in which students are trained to apply their theoretical and practical learning in real-life scenarios for developing and enhancing new design concepts. They are expected to look at human problems from different perspectives. Similarly, design science [27] can be implemented in the five steps as depicted in Figure 1.4.

All these steps are possible only in SLP. Further, in teaching–learning based pedagogy (TLBP) modern digital education is playing a vital role. No doubt the availability of several freeware digital tools has made the education very smart and interactive. This is useful not only to students but also to teachers in explaining the concepts. Various domains of the modern digital world, such as computer-aided design (CAD), robotics, machine learning, deep learning, user interface (UI), and artificial intelligence (AI) [5], have made the young mind very enthusiastic toward learning the modern technologies in a far better way. Also, in the present scenario, various online free educational websites such as *Coursera* [28], *Edx* [29], *Udemy* [30], *NPTEL* [31], *MIT Openware* [32], and *Classcentral* [33] are playing a vital role in enhancing the educational quality of young minds via delivering lectures online in a digital recorded form. The lectures are delivered by highly qualified and recognized SMEs from world-class universities and research centers with the common goal of filling the gap of quality education in various sectors such as engineering, IT, science, art, etc. But there is a common question and misunderstanding that arises in young mind: if everything is available free online or at a very nominal affordable price, then what is the need of teachers? At this stage, the role of teachers becomes very important because in the ocean of knowledge and the available tools, the suitable choice and correct usage of these tools can be interpreted only by a teacher [7].

1.4 TEACHING AND LEARNING PEDAGOGY

Teaching pedagogy is all about transmitting knowledge and necessary skills for extended understanding of subject matter. So, teaching and learning pedagogy (TLP) plays a very significant role in various educational domains such as engineering, management, science, medicine, etc. for developing a functional society. Hence, teaching is the art of propagating knowledge continuously, and that can be achieved via constant practice [34].

From an engineering perspective, *teaching* can be defined as "the ideas of a particular person or group, especially about principles, concept and their application in real life, society and religion". The same idea is transferred to others in order to apply them in realistic scenarios. Hence, "Teaching is about sparking one's inquisitiveness so that one may care about asking critical questions to gain knowledge of something".

FIGURE 1.4 Design science approach toward human problem-solving [27].

1.4.1 TEACHER–STUDENT INTERACTION

"Instructional interactions" are commonly termed as "institutional talk" [35] in which "teachers and students frequently communicate with each other based on their roles of student and teacher" [36]. This communication is done at psychological level and is interpersonal in nature. Teaching is an art that demands a sense of compassion and a caring mindset that helps a teacher in connecting with their students and colleagues [37]. Hence, it can be concluded that *teaching* may be an impressive job for imparting one's knowledge to others [34].

1.4.2 WAYS OF TEACHING AND LEARNING PEDAGOGY

Undoubtedly, the ways of teaching and learning are the two important factors of a healthy and progressive education system. These two factors are dependent on each other for creating a sustainable learning and teaching environment. For a student, it is important to identify their learning style for better outcome-based study. Such identification of one's style of learning helps the teacher in delivering the content and seems to be useful in holding the control of the class. Also, the teacher should continuously work on developing an efficient teaching style for effective teaching experience and for the betterment of students' understanding.

Both teacher and student possess different styles, "but generally, it is supposed that understanding of students' preferences and learning style can be advantageous for both students and teachers and students" [38, 39]. But, along this discussion, one thing is to be noted that students have very creative minds at their age, so their active mind accelerates them to learn new things from various resources. Hence, the adoption of a new learning style is quite difficult for students in the classroom. Hence, the expectation of students coming from various backgrounds is that the teacher should have a customized way of teaching in order to give better insights to students [34].

1.4.3 ENGINEERING ASPECTS OF TEACHING AND LEARNING PEDAGOGY

The learning of students depends on various methods such as audiovisual media, reasoning, memorizing things, reflecting activities, developing models, etc. Therefore, their learning styles vary from time to time and situation to situation. Similarly, teaching styles also vary in such a way that some instructors provide lectures only; others demonstrate lectures by connecting them to real-life situations; some demonstrate using basic background principles and applications; and some emphasize only memorization. Hence, the learning capability of students under these teaching styles could be judged by the compatibility of their learning styles with the teaching styles of teachers. Hence, there exists a mismatch in learning and teaching styles of both engineers and professors. This results in a boring classroom lecture with no outcome, and students get discouraged about the course and curriculum and start learning by various other media apart from professors. "Professors, confronted by low test grades, unresponsive or hostile classes, poor attendance and dropouts, know something is not working; they may become overly critical of their students (making things even worse) or begin to wonder if they are in the right profession. Most seriously, society loses potentially excellent engineers" [40].

Hence, it is necessary that in teaching and learning, both the styles should be customized according to the requirements of learners to bridge the gap of any mismatch occurring in learning and teaching styles [41]. Hence, teaching and learning styles for both teachers and students behave as a supportive mechanism to achieve their goals. This helps in developing a progressive nation as a whole.

1.5 TECHNOLOGY INTEGRATION IN PEDAGOGY

The use of ICT is continuously affecting human life in almost all aspects of today's technological era. The implementation of ICT and OBE in engineering education is a revolutionary approach in reality, but at the same time they possess their own challenges, such as investment of the stakeholders, available resources, etc. [42]. ICT has the potential to help the existing teaching–learning setup by allowing for a quick deployment of "currently practiced activities" that are done in a traditional manner without the use of technology.

In the current technological era, ICT deployment supplements [43] engineering education and comprises "learning management systems" (LMSs), which are actively being utilized globally. The importance of human–computer interaction in easing the teaching–learning process, as well as the use of AI, virtual reality (VR), and other complementary models such as Analysis, Design, Development, Implementation, and Evaluation (ADDIE), technological pedagogical content knowledge (TPACK), and addressing higher cognitive levels are highlighted. In a broader manner, ICT refers to any device or program that can communicate [42, 44, 45]. The successful integration of ICT involves many aspects, such as the "necessary skills, software integration, application, and systems". ICT education consists of four main components that have been delineated in Figure 1.5.

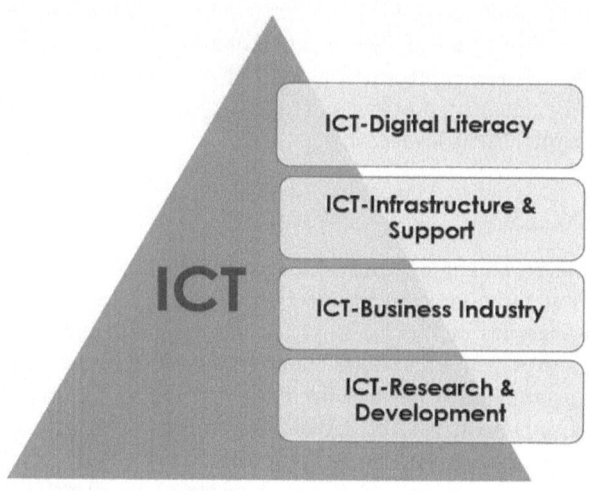

FIGURE 1.5 Schematic of four main components of ICT [43].

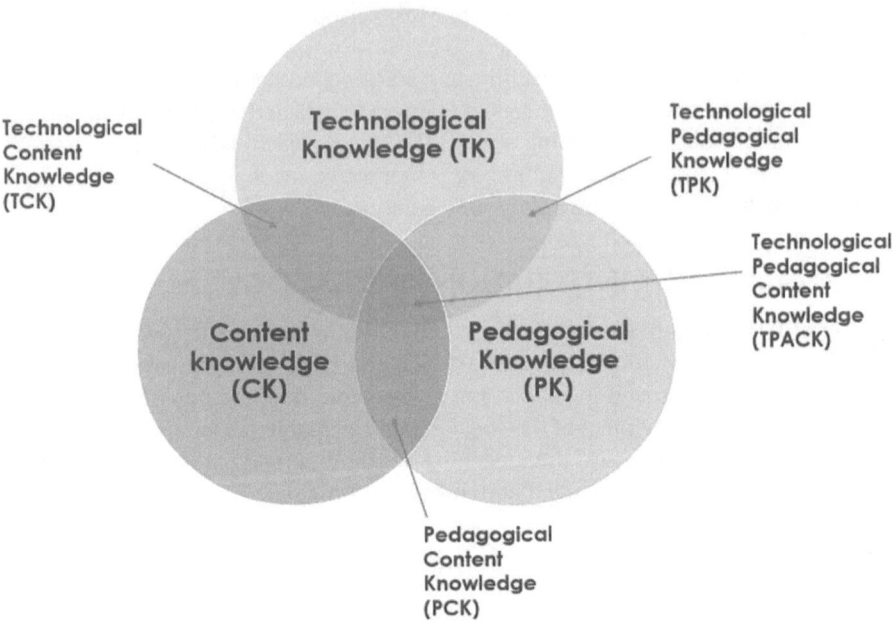

FIGURE 1.6   Schematic Venn diagram showing TPACK Framework [46].

Further, the three factors of the TPACK approach immensely affect the instructor's and student's teaching and learning styles. These three TPACK components provide a fruitful strategy to dealing with a wide range of issues that educators experience while attempting to implement instructional innovation in their classrooms. The TPACK component defines how the educator impacts the created content through various pedagogical methods by applying these three learning approaches, and it must build the foundation for any sustainable educational technology combination. A conceptual overview of TPACK components has been depicted in the Venn diagram in Figure 1.6. Hence, technology will have a positive impact on the teaching, learning, and evaluation processes. This satisfies unique learning needs while also improving pupils' focus levels [46].

1.6   CHALLENGES

Providing outcome-based engineering education is today's requirement toward more sustainable and equitable engineering education. As defined by the US National Academy of Engineering of the National Academies [47], these two issues entail training engineers to conceive and build better goods and services using fewer renewable resources (sustainability) for serving a huge, rising global population (equity). Engineering is a discipline that can be defined as a profession of acquiring the scientific and mathematical knowledge that can be implemented practically in various domains, such as art and culture, social, economic, etc., for obtaining scientific solutions from a technological perspective on a particular issue related to such domains [48].

Several challenges in engineering education have been addressed globally by the researcher's community over the past few decades; the emergence of new technologies has also contributed to redefining the engineering curricula, and the input of educational sciences has suggested the new approaches to TLP. Over the past 100 years, five major shifts in engineering education have been recorded [49].

In regard to the teacher's teaching and learning activity, *teaching practice* is a crucial part. Foncha et al. [50] observed some of the challenges experienced by teachers and students during professional teaching practice, such as "technical–nontechnical resources, placement activities, learner discipline, classroom management", and so on. It also addresses the students' teachers' reflective attitude during their exposure to the learning environment in order to acquire theoretical features. Various necessities in today's technological era are challenges in actually handling the teaching process. Some of the classroom-related challenges are as follows [51]:

- Course/syllabus revision is required to meet today's demand of industry and society.
- Planning is necessary while updating the course contents and focusing on developing an innovative education system. It may consume much time to develop such a system.
- Developing a good statistical report related to teaching requires a lot of paper work and data collection, which itself is a challenge.
- The teacher must be capable of exploring various teaching styles while having interactions with every new batch of students in a new session.
- Educational family background also plays a significant role in a progressive career of a student. It is not uncommon for a student to be the first in his/her family to receive such an education. As a result, it is also necessary to hold counselling sessions and mentor the student.

Further, it is well-known that in today's era pedagogy-based education system has itself become a challenge. The demand for engineers has never grown as drastically as today. Peter F. Drucker has most significantly observed the challenges of the 21st century. According to him, "The most important and indeed the truly unique contribution of management in the 20th Century was the fifty-fold increase in the productivity of the MANUAL WORKER in manufacturing. The most important contribution management needs to make in the 21st Century is similarly to increase the productivity of KNOWLEDGE WORK and of the *KNOWLEDGE WORKER*". We have a more precise dimension of the issues we confront, especially in engineering education, if we substitute "*KNOWLEDGE WORK*" and "*KNOWLEDGE WORKER*" with *EDUCATION* and *EDUCATOR* (or instructor).

"Many students' lives today are filled with technology that

- gives them mobile access to information and resources 24/7,
- enables them to create multimedia content and share it with the world, and
- allows them to participate in online social networks where people from all over the world share ideas, collaborate, and learn new things.

Outside school, students are free to pursue their passions in their own way and at their own pace. The opportunities are limitless, borderless, and instantaneous" [48].

1.7   CASE STUDY

The entire education system has suddenly been shaken due to sudden spread of COVID-19. The higher education system (HES) has been affected drastically due to massive lockdowns around the globe. Fear of COVID-19 has attacked people at their psychological level [52]. The regular trend of the HES was helpless and forced to adopt a new trend of online education and transform the whole setup. However, a positive aspect of such online learning for students was to focus on the productive use of their time to avoid any loss of study and semester or year [53].

During the pandemic, the pedagogy of online education system (POES) has been regarded as a boon for supporting the continuity of education and forcing a rethink of the entire policy and planning of higher education institutions (HEIs) for both students as well as the teachers and staff [54–56]. The new pedagogy of online education and the adoption of digital education are reflected in various forms in today's teaching and learning environment. The very recent trends have helped the students and teachers in the following ways:

- comprehensive online classes
- online assessments in digital mode
- personalization of education
- developing a new kind of examination management

1.8   CHALLENGES FACED DURING COVID-19

1.8.1   Moving from Classroom to Online Mode

The introduction of the online mode stimulated the interest of many teachers and students throughout the world. Teachers had already begun developing presentations for delivery of courses through the internet. Before the COVID-19 pandemic, several universities offered online courses. Many academic and staff members had received training in the use of online learning systems, although this online option may be challenging for some faculty because they are not technologically literate [57–59].

However, this sudden shift to the online mode was quite a challenging task for both students and teachers which affected both groups in different ways at the beginning. This sudden shift has made a complete system breakdown and the practical education reached almost zero level. This has increased the stress level of both students and teachers without providing any outcome-based education system. Many surveys have reported the mental conditions during the pedagogy of online teaching and learning. Figure 1.7 shows the diverse responses of students toward the sudden shifting from classroom to online mode. Students were asked to choose one of the following options to describe how they felt at the time of the abrupt online journey of education: *excited, stressed, anxious, or not sure* [60].

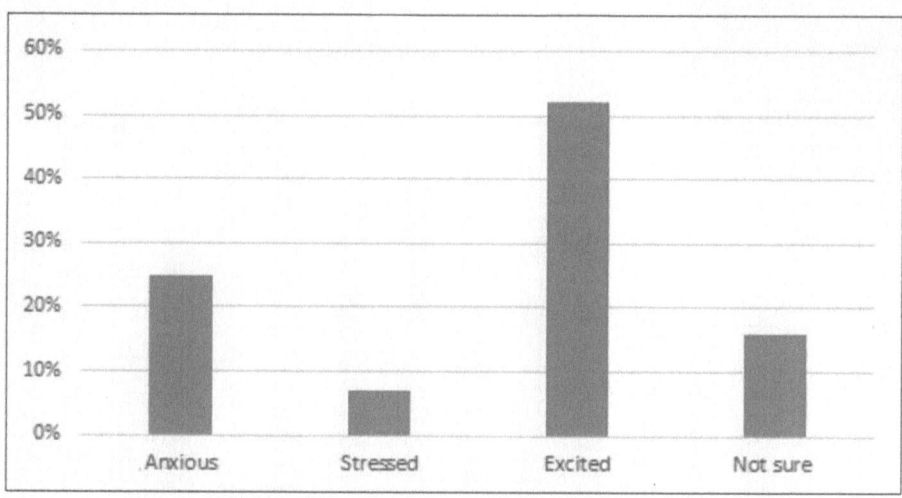

FIGURE 1.7 Student's response toward the shifting from classroom mode to online mode [60].

1.8.2 MOVING FROM CLASSROOM ASSESSMENT TO ONLINE ASSESSMENT

During COVID-19, most institutions cancelled or postponed their semester examinations, but in the meantime, the teachers and students were helped out of the dilemma by an initiative taken from LMS in the form of providing an option of digital education. Including online assessments in lessons prepared for classroom instruction was a difficult issue. Both students and teachers were confused about how assignments, projects, and other continuing learning activities would be assessed. Students who did not have access to the internet were negatively impacted and faced difficulty in completing assessments, which had a bad influence on their grades [61]. Figure 1.8 shows survey results based on various challenges faced by the instructors during the COVID-19 situation.

Many students studying abroad found it difficult to return to their home country in pandemic circumstances [62]. Providing food, shelter, and security for these international students proved to be a difficult task for authorities. Students also needed to be adequately protected from any contact with infected individuals and had to remain in self-isolation until the situation returned to normal. Extending one's stay owing to exam delays can create a financial bind. COVID-19's disruption may have an influence on international student admissions for the coming academic year [63].

During the pandemic, various surveys were carried out regarding teaching and learning methodologies of the instructors and students. The students' qualitative responses were analyzed and grouped into three main categories: *pedagogical reasons*, *psychological reasons*, and *technological reasons*.

Instructors' efforts in arranging course material, controlling class time, generating given tasks, making sessions interactive, and ensuring that students attained the same learning outcomes as in face-to-face delivery were among the *pedagogical aspects*. Instructors who were "capable users of technology" or "being creative and innovative with the use of diverse tools to engage the students online" were

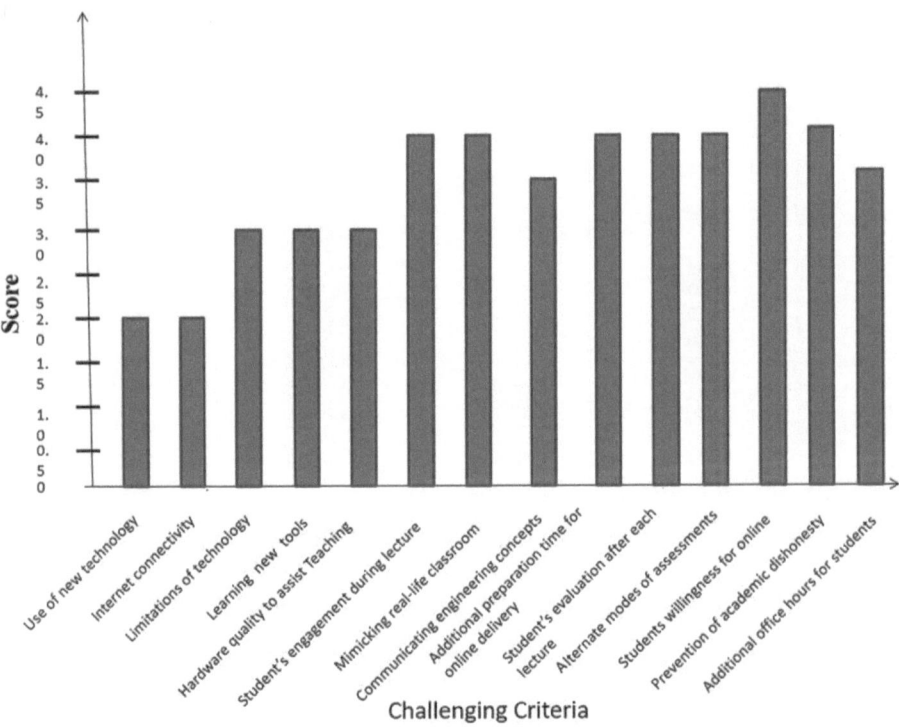

FIGURE 1.8 Challenges faced by instructors in various domains during COVID-19 [60].

praised for technological considerations. These were considered under *technological aspects*. The *psychological aspects* appeared to be mostly related to the teachers' attitudes, such as supportive behavior, flexibility, friendliness, shared sympathy for the students, and understanding and adjusting the requirements of the students. These three aspects of findings have been delineated in Figure 1.9.

1.9 CONCLUSION

Pedagogical aspects of teaching and learning methods in engineering education are widely acclaimed. Design pedagogy in engineering design has introduced tremendous motivation in developing new product design and new innovation. LBD and DBL approaches have provided a pathway toward developing DBP in the domain of higher education. The instructional structure of such pedagogical approaches has developed a progressive and creative mindset in students and enhanced their problem-solving skills. Further, the integration of technology in TLP has affected the ways of engineering education in multiple ways. Such integration has come up with various positive aspects of utilization of ICT in a far better way and has furthered outcome-based education in the light of future pedagogical approaches. An extensive study in the pedagogy of outcome-based engineering education by the authors has provided a modest contribution to this current work. The current work highlights the

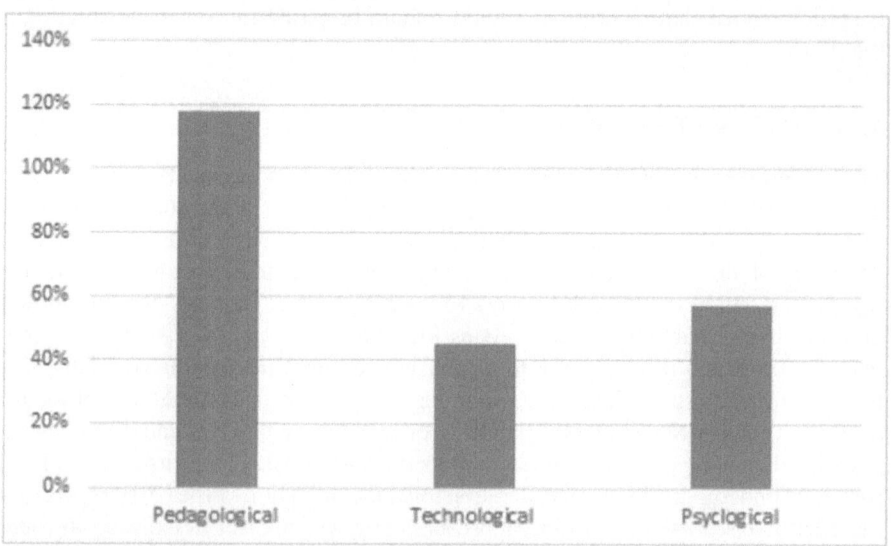

FIGURE 1.9 Responses for preferring the instructors [60].

incorporation of a pedagogical approach in engineering education. SLP is another aspect of design pedagogy in which students are taught to implement their cognitive thoughts and learning process in real-world problems by moving from a theoretical to a practical approach. Further, the various components of the TPACK framework have helped a lot in building a strong foundation in the teaching–learning approach.

Further, the current study suggests various modern pedagogical tools in today's technological era that have supported the education system at the time of COVID-19 crisis. The present study also discusses the pedagogical approach and integration of technology in teaching and learning-based education through a case study. The case study highlights the sudden shift from classroom mode to online teaching–learning. Despite the availability of technology and applicable platforms for synchronous and asynchronous modes of delivery, many academic institutions were unprepared for the abrupt online migration. However, to ensure continuity in the teaching–learning process, this study contends that the adoption of new learning styles and theories may teach us a lot about how to use technology as a tool for effective pedagogies and learning strategies. During the abrupt transfer online during COVID -19, students experienced technological, psychological, and educational hurdles, according to the report. Due to a lack of preparedness and the faculty's inability to adjust well to the COVID emergency, it was obvious that these issues had an impact on the students' learning process. Because they were not fully prepared for the online delivery and showed anxiousness while teaching and learning, instructors faced challenges that were plainly obvious to the students. Also, it is evident that the offline mode of education can never be replaced by online mode.

Despite teachers' unpreparedness and the challenges of online delivery, students praised the instructors' efforts but expressed dissatisfaction with aspects such as the online learning course, online assessments, and online examinations. Therefore, it is recommended that institutions should remain prepared for any such situation in the future.

1.10 FUTURE TRENDS

The student-centered OBE approach in higher education engineering programs is quite beneficial because it focuses on quality of achievement and student capability. However, various barriers and challenges need to be tackled in order to enhance the pedagogy of TLP. The mismatch occurring between the teachers and students during TLP is one of the biggest barriers to learning. Hence, there is need to develop practical-based teaching–learning styles for better outcome. A practical teaching–learning process will also serve to establish a resourceful sustainable relationship between instructors and learners. Technology integration is a crucial part of today's learning, along with the careful handling of technology barriers and removing the associated challenges (as discussed in preceding sections). This is itself a challenge that must undergo a critical consideration toward eradication of the barriers. Further, the continuous practical learning approach along with industrial exposure through various seminars and workshops on engineering topics will serve a lot toward developing OBE. Research-based teaching styles and learnings will stimulate the learners more in developing research studies with new insights and thoughts toward new and innovative product and process development.

REFERENCES

1. S. Kersten, Approaches of Engineering Pedagogy to Improve the Quality of Teaching in Engineering Education, 2018, pp. 129–139, https://doi.org/10.1007/978-3-319-73093-6_14
2. S. Kersten, H. Simmert, and D. Gormaz, Engineering Pedagogy at Universities in Chile—A Research and Further Education Project of TU Dresden and Universidad Autónoma de Chile, in Expanding Learning Scenarios. EDEN Conference Barcelona, 2015, http://wwwpub.zih.tudresden.de/~kersten/Artikel/Engineering%20Pedagogy_EDEN%202015.pdf
3. C. Lavanya, J. N. Murthy, and S. Kosaraju, Assessment Practices in Outcome-Based Education, in Methodologies and Outcomes of Engineering and Technological Pedagogy, IGI Global, Hershey PA, 2020, pp. 50–61, https://doi.org/10.4018/978-1-7998-2245-5.ch004
4. H. Poser, On Structural Differences Between Science and Engineering, Society for Philosophy and Technology Quarterly Electronic Journal, 4(2), 1998, pp. 128–135, https://doi.org/10.5840/techne1998426
5. R. Kumar and K. Kumar, Intelligence-Assisted Cobots in Smart Manufacturing, in Advanced Computational Methods in Mechanical and Materials Engineering, CRC Press, Boca Raton, 2021, pp. 19–41.
6. T. Swirski, Unleashing the Imagination in Learning, Teaching and Assessment: Design Perspectives, Innovative Practices and Meaning Making, Paper presented at the ATN Assessment Conference, Australia: Australian Technology Network, 2010, https://www.uts.edu.au/sites/default/files/Swirski.pdf
7. J. Srinivas, Modern Pedagogy Tools in Engineering Education, Chapter 6, in Methodologies and Outcomes of Engineering and Technological Pedagogy, IGI Global, Hershey PA, 2020, pp. 78–86, https://doi.org/10.4018/978-1-7998-2245-5.ch006

8. G. Wilson, Technical Communication and Late Capitalism, Journal of Business and Technical Communication, 15(1), 2001, pp. 72–99, https://doi.org/10.1177/10506519010 1500104

9. L. Melonc, et al. A Field-Wide Metasynthesis of Pedagogical Research in Technical and Professional Communication, Journal of Technical Writing and Communication 50(1), 2019, pp. 91–118. https://doi.org/10.1177/0047281619853258

10. K. Kumar and J. P. Davim, Methodologies and Outcomes of Engineering and Technological Pedagogy, vol. 1, IGI Global, Hershey PA, 2020, https://doi.org/10.4018/978-1-7998-2245-5

11. R. Scherer, et al. The Importance of Attitudes Toward Technology for Pre-Service Teachers' Technological, Pedagogical, and Content Knowledge: Comparing Structural Equation Modeling Approaches, Computers in Human Behavior, 80, 2017, https://doi.org/10.1016/j.chb.2017.11.003

12. Web of Science, https://www.webofscience.com/wos/woscc/analyze-results/c7242bf8-9acc-45f3-8ad7-619e16438337-2254f1ae (accessed Feb. 08, 2022)

13. Scopus.com, https://www.scopus.com/results/results.uri?sort=plf-f&src=s&st1=engine ering+pedagogy&sid=ee8d2fe76f37b5616cb3490fe788cbbc&sot=b&sdt=b&sl=35&s= TITLE-ABS-KEY%28engineering+pedagogy%29&origin=searchbasic&editSaveSea rch=&yearFrom=Before+1960&yearTo=Present (accessed Feb. 08, 2022)

14. A. Royalty, Design-Based Pedagogy: Investigating an Emerging Approach to Teaching Design to Non-Designers, Mechanism and Machine Theory, 125, 2018, pp. 137–145, https://doi.org/10.1016/j.mechmachtheory.2017.12.014

15. H. A. Simon, The Sciences of the Artificial, Third edition, The MIT Press, Cambridge, MA, 2019, https://mitpress.mit.edu/books/sciences-artificial

16. C. L. Dym, A. M. Agogino, O. Eris, D. D. Frey, and L. J. Leifer, Engineering Design Thinking, Teaching, and Learning, Journal of Engineering Education, 94(1), 2005, pp. 103–120, https://doi.org/10.1002/j.2168-9830.2005.tb00832.x

17. N. Cross, Engineering Design Methods: Strategies for Product Design, Fourth Edition, John Wiley & Sons, England, 2000, https://www.wiley.com/en-us/Engineering+Design+ Methods%3A+Strategies+for+Product+Design%2C+4th+Edition-p-9780470519264

18. Y. Doppelt, M. M. Mehalik, C. D. Schunn, E. Silk, and D. Krysinski, Engagement and Achievements: A Case Study of Design-Based Learning in a Science Context, Journal of Technology Education, 19(2), 2008, pp. 22–39, http://hdl.handle.net/10919/8390

19. G. Pahl, W. Beitz, J. Feldhusen, and K.-H. Grote, Engineering Design: A Systematic Approach, Springer, London, 2007, https://doi.org/10.1007/978-1-84628-319-2

20. H. A. Simon, The Science of Design: Creating the Artificial, in The Sciences of the Artificial, MIT Press, Cambridge MA, 1996, pp. 111–138, https://ieeexplore.ieee.org/document/6282803

21. M. Korn and R. E. Silverman, Forget B-School, D-School Is Hot, The Wall Street Journal, https://www.wsj.com/articles/SB10001424052702303506404577446832178537716

22. N. Cross, Design Thinking: Understanding How Designers Think and Work. Berg Publishers, 2011, https://doi.org/10.5040/9781474293884

23. J. L. Kolodner, P. J. Camp, D. Crismond, B. Fasse, J. Gray, J. Holbrook, S. Puntambekar, and M. Ryan, Problem-Based Learning Meets Case-Based Reasoning in the Middle-School Science Classroom: Putting Learning by Design Into Practice, Journal of the Learning Science, 12(4), 2003, pp. 495–547, https://doi.org/10.1207/S15327809JLS1204_2

24. J. Herrington and R. Oliver, An Instructional Design Framework for Authentic Learning Environments, Educational Technology Research and Development, 48(3), 2000, pp. 23–48, https://doi.org/10.1007/BF02319856

25. J. R. Savery and T. M. Duffy, Problem Based Learning: An Instructional Model and Its Constructivist Framework, Educational Technology, 35(5), 1995, pp. 31–38, https://www.jstor.org/stable/44428296

26. E. Felix and M. Brown, The Case for a Learning Space Performance Rating System, Journal of Learning Spaces, 1(1), 2011, https://libjournal.uncg.edu/index.php/jls/article/viewArticle/287/154

27. K. Peffers, T. Tuunanen, M. A. Rothenberger, and S. Chatterjee, A Design Science Research Methodology for Information Systems Research, Journal of Management Information Systems, 24(3), 2014, pp. 45–77, https://doi.org/10.2753/MIS0742-1222240302

28. coursera.org, https://www.coursera.org/courses?query=pedagogy (accessed Feb. 10, 2022)

29. edx.org, https://www.edx.org/about/research-pedagogy, (accessed Feb. 10, 2022)

30. udemy.com, https://www.udemy.com/courses/search/?src=ukw&q=pedagogy (accessed Feb. 10, 2022)

31. National Programme on Technology Enhanced Learning (NPTEL), https://nptel.ac.in/courses/121/105/121105010/ (accessed Feb. 10, 2022)

32. ocw.mit.edu, https://ocw.mit.edu/educator/pedagogy-highlights/ (accessed Feb. 10, 2022)

33. classcentral.com, https://www.classcentral.com/search?q=pedagogy (accessed Feb. 10, 2022)

34. A. Dixit, K. K. Saxena, and J. K. Dixit, Teaching and Learning: Pedagogy With Purpose, in Methodologies and Outcomes of Engineering and Technological Pedagogy, edited by Kaushik Kumar and J. Paulo Davim, IGI Global, Hershey PA, 2020, pp. 20–27. https://doi.org/10.4018/978-1-7998-2245-5.ch002

35. J. Heritage and L. Angeles, Conversation Analysis and Institutional Talk : Analyzing Distinctive Turn-Taking Systems, in Dialoganalyse VI/2, edited by Svetla Cmejrkova, Jana Hoffmannová, Olga Müllerová and Jindra Svetlá, Max Niemeyer Verlag, Berlin, Boston, 2017, pp. 3–18, https://doi.org/10.1515/9783110965049-001

36. N. D. Dobransky and A. B. Frymier, Developing Teacher-Student Relationships Through Out of Class Communication, Communication Quarterly, 52(3), 2004, pp. 211–223, https://doi.org/10.1080/01463370409370193

37. E. Taylor, Use of Non-Situational Identities in Teacher-Student Interaction, Linguistics and Education, 66, 2021, 100997, https://doi.org/10.1016/j.linged.2021.100997

38. K. O'Brien, Global Environmental Change III: Closing the Gap Between Knowledge and Action, Progress in Human Geography, 37(4), 2013, pp. 587–596, https://doi.org/10.1177/0309132512469589

39. P. Courtenay-Hall and L. Rogers, Gaps in Mind: Problems in Environmental Knowledge-Behaviour Modelling Research, Environmental Education Research, 8(3), pp. 283–297, https://doi.org/10.1080/13504620220145438

40. R. Felder and L. Silverman, Learning and Teaching Styles in Engineering Education, Engineering Education, 78(7), 1988, pp. 674–681.

41. Y. Sipos, B. Battisti, and K. Grimm, Achieving Transformative Sustainability Learning: Engaging Head, Hands and Heart, International Journal of Sustainability in Higher Education, 9(1), 2008, pp. 68–86, https://doi.org/10.1108/14676370810842193

42. P. Gupta, T. Kulkarni, V. Barot, and B. Toksha, Applications of ICT: Pathway to Outcome-Based Education in Engineering and Technology Curriculum, in Technology and Tools in Engineering Education, 2021, pp. 109–142, CRC Press, Boca Raton, https://doi.org/10.1201/9781003102298-7

43. A. Chen, J. Castillo, and K. Ligon, Information and Communication Technologies (ICT): Components, Dimensions, and its Correlates, Journal of International Technology and Information Management, 24(4), 2015, pp. 24–46, https://scholarworks.lib.csusb.edu/jitim/vol24/iss4/2/

44. F. Schiliro and K.-K. R. Choo, The Role of Mobile Devices in Enhancing the Policing System to Improve Efficiency and Effectiveness, Chapter 5, in Mobile Security and Privacy, Elsevier, 2017, pp. 85–99, https://doi.org/10.1016/B978-0-12-804629-6.00005-5

45. X. Hu, X. Hou, C.-U. Lei, C. Yang, and J. Ng, An Outcome-Based Dashboard for Moodle and Open edX, in Proceedings of the Seventh International Learning Analytics & Knowledge Conference, Mar. 2017, pp. 604–605, https://doi.org/10.1145/3027385.3029483

46. D. P. Garapati and S. M. Padmaja, Technology in Engineering Pedagogy to Progress the Excellence of Teaching: Teaching Learning Process, in Methodologies and Outcomes of Engineering and Technological Pedagogy, IGI Global, Hershey PA, 2020, pp. 1–19, https://www.doi.org/10.4018/978-1-7998-2245-5.ch001

47. US National Academy of Engineering of the National Academies, NAE Grand Challenges for Engineering, National Academy of Engineering, 2008. http://www.engineeringchallenges.org/challenges.aspx (accessed Feb. 14, 2022)

48. M. E. Auer and D. Zutin, New Pedagogic Challenges in Engineering Education, 2011 Frontiers in Education Conference (FIE), 2012, pp. 1–7, https://www.doi.org/10.1109/FIE.2011.6142845

49. J. E. Froyd, P. C. Wankat, and K. A. Smith, Five Major Shifts in 100 Years of Engineering Education, Proceedings of IEEE, 100(Special Centennial Issue), 2012, pp. 1344–1360, https://www.doi.org/10.1109/JPROC.2012.2190167

50. J. W. Foncha, J.-F. A. Abongdia, and E. O. Adu, Challenges Encountered by Student Teachers in Teaching English Language during Teaching Practice in East London, South Africa, International Journal of Educational Sciences, 9(2), 2015, pp. 127–134, https://doi.org/10.1080/09751122.2015.11890302

51. J. Aher, S. Desai, H. Pande, and A. Thengade, Classroom Practices: Issues, Challenges, and Solutions, in Technology and Tools in Engineering Education, CRC Press, Boca Raton, 2021, pp. 29–40, https://doi.org/10.1201/9781003102298-2

52. U. G. Singh and D. K. Sharma, Investigating Academic Transition during the COVID-19 Pandemic, in Technology and Tools in Engineering Education, CRC Press, Boca Raton, 2021, pp. 59–79, https://doi.org/10.1201/9781003102298-4

53. K. Mugo, N. Odera, and M. Wachira, Surveying the impact of COVID-19 on Africa's higher education and research sector, 2020, https://www.africaportal.org/features/surveying-impact-covid-19-africas-higher-education-and-research-sectors/ (accessed Feb. 15, 2022)

54. Cape Argus, This is how SA varsities are implementing online teaching amid Covid-19 lockdown, https://www.iol.co.za/news/south-africa/western-cape/this-is-how-sa-varsities-are-implementing-online-teaching-amid-covid-19-lockdown-46930129 (accessed Feb. 15, 2022)

55. USAF, Online learning is integral to the future of higher education; embrace it or become irrelevant, https://www.usaf.ac.za/online-learning-is-integral-to-the-future-of-higher-education-embrace-it-or-become-irrelevant/#:~:text=USAf 0 comment-,Online learning is integral to the future of Higher,embrace it or become irrelevant&text=%22 (accessed Feb. 15, 2022)

56. M. Yamin, Counting the Cost of COVID-19, International Journal of Information Technology, 12(2), 2020, pp. 311–317, https://doi.org/10.1007/s41870-020-00466-0

57. M. Lim, Educating despite the Covid-19 outbreak: lessons from Singapore, 2020, https://www.timeshighereducation.com/blog/educating-despite-covid-19-outbreak-lessons-singapore (accessed Feb. 16, 2022)

58. U. G. Singh, Academic digital literacy – A journey we all need to take, 2020, https://www.universityworldnews.com/post.php?story=20200630085507410

59. P. Sahu, Closure of Universities Due to Coronavirus Disease 2019 (COVID-19): Impact on Education and Mental Health of Students and Academic Staff, Cureus, 12(4), 2020, e7541. doi: 10.7759/cureus.7541.

60. V. Ahmed and A. Opoku, Technology Supported Learning and Pedagogy in Times of Crisis: The Case of COVID-19 Pandemic, Education and Information Technologies, 27(1), 2022, pp. 365–405, https://doi.org/10.1007/s10639-021-10706-w

61. E. Dill, K. Fischer, B. McMurtrie, and B. Supiano, As Coronavirus Spreads, the Decision to Move Classes Online Is the First Step. What Comes Next? 2020. https://www.chronicle.com/article/as-coronavirus-spreads-the-decision-to-move-classes-online-is-the-first-step-what-comes-next/ (accessed Feb. 16, 2022)

62. E. Bothwell, Flexible Admissions Could Mitigate Covid-19 Impact, 2020, https://www.timeshighereducation.com/news/flexible-admissions-could-mitigate-covid-19-impact#:~:text=Experts%20say%20institutions%20should%20allow,date%20and%20relax%20entry%20requirements&text=Universities%20should%20focus%20on%20introducing,coronavirus%20crisis%2C%20according%20to%20experts.
63. S. Timmis, P. Broadfoot, R. Sutherland, and A. Oldfield, Rethinking Assessment in a Digital Age: Opportunities, Challenges and Risks, British Journal of Educational Technology, 42(3), Jun. 2016, pp. 454–476, https://doi.org/10.1002/berj.3215

2 Life-Long Learning Practices

A Conceptual Analysis

A. Udhaya Kumar
Hindusthan College of Engineering and Technology
Coimbatore, Tamil Nadu, India

A. Kishore Kumar
Sri Ramakrishna Engineering College
Coimbatore, Tamil Nadu, India

CONTENTS

2.1 INTRODUCTION

Transitions are a time when things are changing. The readiness of people to adapt and learn is a critical issue. Because people are locked in their routines, opportunities for change may go unnoticed. People who are accustomed to adapting will rely on previous transactions, making little alterations as needed. People who seek new information, want to attempt new things, and are aware of their surroundings' demands and obstacles have learnt how to be generative. Some have had the chance to play a pivotal role in bringing about paradigm-shifting transformation.

International organizations have emphasized the importance of lifelong learning for active voters in the information economy for many years. In the context of economic development and social advancement, the United Nations Educational, Scientific, and Cultural Organization (UNESCO) popularized the term "lifelong

DOI: 10.1201/9781003083160-3

learning" [1, 2]. The European Union selected 1996 the "Year of Womb-to-Tomb Learning" [3], in line with the Organization for Economic Cooperation and Progress (OECD) [4], because womb-to-tomb learning is crucial for economic development and equity, particularly in developing countries. Universities may help students develop fundamental learning measures by doing research, training lecturers to believe in the interconnectedness of long learning and serving as role models, and creating learning environments that encourage students to review throughout their lives.

Higher education program goals embrace womb-to-tomb learning competencies. Womb-to-tomb learning is a definite component of the aims within the European Higher Education Area (EHEA), with all university graduates required to "have mastered the learning skills necessary for them to still conduct additional studies with a high degree of autonomy" [5]. The United States of America's (USA) certification authority for engineering and technology, ABET [6], adds a demand that engineering graduates have "a recognition of the necessity for, and talent for, interaction in womb-to-tomb learning."

2.2 LIFELONG LEARNING COMPETENCIES

"Intentional learning is a type of learning that people do throughout their lives in order to achieve personal and professional fulfilment and improve their quality of life." To achieve personal and professional accomplishment and to enhance the quality of one's life, people actively pursue lifelong learning throughout their lives [7, 8]. The basic possibilities of LLL are reflected in its name: "lifelong" from conception to death and "learning" rather than "instruction," with learners responsible for their own learning processes [9]. Educators play a critical part in the advanced approach of lifelong learning growth. Their pedagogical choices can have a significant impact on how students learn in the long run. Competence-based education demands the development of a learning setting within which info is transferred in eventualities that enable students to mobilize, combine, and transcend their previous information. Domain-specific data underpins competencies, which will be used in any gift or future observation. As a result, rather than being guided by tradition or authoritative viewpoints, learning should be guided by the strain of observation. Competency-based learning is outlined by the exchange of data from learning to observe, discourse learning, and silent knowing. [10]

All professionals share key competencies, also known as generic, transdisciplinary, or crosswise competencies, which encompass communication, teamwork, and data technology attainment; explicit competencies, which can encompass concrete "technical" data in a very specific space, align with the two main styles of competencies. The need to keep up with change has never been more important than it is today, thanks to the growing amount and changing type of technical information.

2.3 ENGINEERING COMPETENCES FOR LIFELONG LEARNING

Learners should be aware that their employability in today's world depends on their capacity to adapt to a number of different job needs as they progress from academia to industry. Those who most effectively show this ability to be educated for the remainder of their lives seem to be the most prosperous. In today's business, workers' are expected to point out initiative, independence, and demanding thinking skills. Competence requires the ability to adapt and respond to new situations throughout time, which involves lifelong learning [11]. Skilled competence is defined as having the information, skills, attitude, and values needed to practice a profession, along with the ability to speak effectively. The ability to solve professional problems autonomously and adaptively, as well as to collaborate in a skilled environment, using and communicating knowledge to others in engineering, a complex field that is always changing due to technological advancements, emphasizes the significance of lifelong learning. Meredith recommends creating a personal (and professional) development plan (PDP) early in one's employment to ensure that one can learn for the rest of one's life. This plan should be revised on an everyday basis and used as a guide for reaching skilled and private objectives. Meredith [12] explains the process in the following ways:

1. Evaluate your data, skills, and social network in light of the requirements of your new employment.
2. Identify critical gaps in your professional expertise that you should fill in the early stages of your career.
3. Make a to-do list with deadlines for learning and development activities.
4. Track your progress on a daily basis to guarantee that you continue to improve in your career.
5. Once you've changed your schedule, focus on basic generic competencies as well as specialized and technical training.

2.4 OUTCOME-BASED EDUCATION

In today's education industry, outcome-based education (OBE) and the accreditation process are becoming increasingly important in guiding students' professional careers. The quality metrics of the education system are the teaching–learning process, pedagogic principles, accreditation, outcome-based curriculum, and assessment. Graduate qualities are measured through direct and indirect achievements, which help to improve the teaching–learning process and students' performance and to bridge the gap between active teaching and outcome-based learning. The low employability of recent engineering graduates, because of the absence of premier engineering institutions in worldwide rankings, attests to the poor quality of engineering education in Asian countries. It's becoming progressively evident that raising the quality of engineering education in Asian countries is not possible until "quality" is outlined in terms of "learning outcomes" instead of grades and then measured consequently. The Washington Accord (WA), which was formed to improve worldwide quality in Engineering Education, places a premium on training results. Asian countries have acted as a test case for the WA since 2007. Despite the fact that the University Grants Commission (UGC) regulation on mandatory certification of higher education institutions (HEIs) from 2012 emphasizes the importance of achieving learning objectives, it leaves the determination to the HEIs. Adopting outcomes-based education might benefit all stakeholders – students, parents, engineering institutions, businesses, and the government – not just in terms of increasing graduate engineers' employability, but also in terms of gaining access to large international prospects.

2.4.1 ENGINEERING EDUCATION

Expectations engineering has a direct and significant impact on people's lives, economic development, and, as a result, the availability of welfare work. Engineering education should make sure that students graduate with the skills they need to succeed as professionals because engineers have high expectations and, as a result, the capacity to address complex social challenges. A strong mathematics and scientific basis, as well as training in specialized engineering specialties, should be included in this curriculum. Because style is such an important talent for engineers, students should conquer increasingly complex problems as they go through the academic approach. The complexity of today's engineering problems needs a strong foundation in areas such as social science, communication, teamwork, and, as a result, the gift global governance environment. Engineers need to conduct themselves in a skilled manner that's honest, objective, and fair.

Engineering graduates should be employable as well as qualified to pursue a second degree in engineering if they so desire. Contemporary graduates who add core engineering vocations are frequently involved in duties such as product design, production, sale, service, and maintenance, and they should have the necessary abilities to accomplish the assigned roles. Engineers' certificates become more mobile as a result of the economic process, making it easier to deploy technical abilities where they're needed. Qualification also permits students to pursue educational activity in any location they need. Potential employers assess contemporary engineering graduates on the basis of the abilities and competency needed on the work, instead of on the information learned throughout their study. Learning outcomes offer verifiable descriptions of what learners are expected to understand and/or be ready to perform. Rather that specializing in the teacher's goals, the coaching-outcomes approach concentrates on what the learner has accomplished and can demonstrate at the conclusion of the coaching session. The objective of a teaching activity differs from the learning outcome because of this student-centered approach. The program's instructional objectives are written from the perspective of the instructor, and they address the expected outcomes of teaching and learning. Learning outcomes, on the other hand, contemplate learning from the attitude of the scholars and concentrate on the accomplished/shown results.

The necessary aspects of outcomes-based education are program educational objectives (PEOs), program outcomes (POs), and course outcomes (COs). The program's instructional objectives are broad statements that replicate the program graduates' skills and ambitions. Program outcomes are statements that describe what students ought to perceive and be able to do at the conclusion of the program. When finishing a course, students are expected to grasp and be able to do certain things. PEOs, POs, and COs ought to all work together to attain identical goals. Once we begin formulating learning objectives, we'll notice that different tasks need different levels of information and ability; some need only memorization, whereas the bulk need sharpened analytical skills and inventive thinking. There were designed as domains for psychological options (thinking and problem-solving skills), emotional aspects (attitudes, price systems), and content. Figure 2.1 depicts a concept map on constructive alignment of assumed outcomes, delivery, and assessment.

The most vital needs for any outcomes-based qualification are a transparent understanding of the program's goals and objectives, teaching methods that support the event

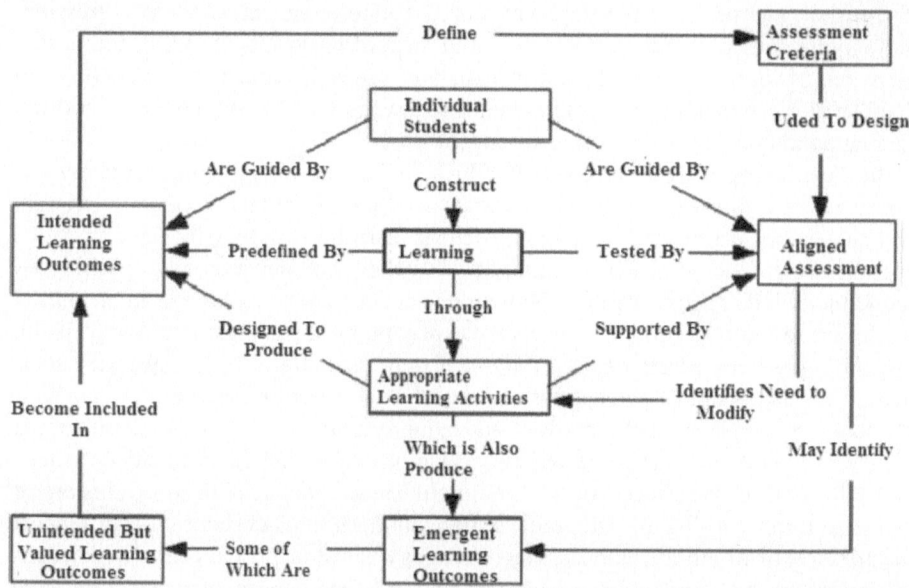

FIGURE 2.1 Constructive alignment of intended outcomes, delivery, and assessment concept map. (Prof. O. R. S. Rao "Outcome Based Engineering Education – Need Of The Hour" *The Journal of Engineering Education* July, 2013.)

of the required competencies, and assessment procedures that reliably monitor whether or not the established targets are being met. As a result, the establishment ought to make sure that its programs offer a unified assortment of discipline-specific and complementary info areas, with integration of essential skills and values. Furthermore, as the student progresses through the program, sufficient opportunities for the demonstration and assessment demonstration of required competencies should be provided to guarantee that students are well-prepared for competent observation and lengthy learning.

Self-directed and informal learning are core talents for effective womb-to-tomb learning, each of which are difficult to show and assess; however, they may be offered effectively through a mixture of info and co-curricular efforts. Educators should specialize in approaches to develop this holistic talent that don't seem to be unnatural to specific courses or topics of study so as to develop students to be effective at informal learning once they are practicing professionals. In addition, enabling children to become self-reliant learners needs specific care. In international data societies, one should keep up one's data within the dynamic world around North American countries. In such cultures, the general populace should be exposed to up-to-date information, skills, and vocational training that can help them within the geographical area. Womb-to-tomb learning is defined as learning that extends beyond traditional schooling and into maturity. It encompasses formal, non-formal, and informal learning that is flexible, broad, and available to people of all ages in a number of situations. Learning is said to be age and classroom agnostic, occurring at all stages of life and in a variety of settings. It encourages not only social participation, active citizenship, and personal development, but also autonomy, antagonistic aggression, and employability. Womb-to-tomb learning, according to UNESCO, encompasses all types of learning (formal, non-formal, and informal) and ages (from birth to death). Two groundbreaking studies released by the United Nations described the fundamental concepts of womb-to-tomb learning.

In the expanding worldwide economy, where the knowledge-based economy has overtaken past kinds of economic research in terms of national economic strength and subject socio-economic direction, lifelong learning is especially important. It's become a big concept in the way individuals believe in education and coaching all round the world. There's little doubt that universities play a crucial role in encouraging womb-to-tomb learning that meets the expectations of students. The World Health Organizations sees education as a force that may facilitate them in reworking their lives. It's been noted that perceptions of womb-to-tomb learning vary greatly, not simply between countries but also between education system sub-sectors. As a result of the numbers of youth in Asian countries such as China and others, the difficulty of womb-to-tomb learning also affects young children. India has a long and illustrious history of womb-to-tomb learning. In truth, India's civilization, culture, and education all support the principle of womb-to-tomb learning. Pedagogy has three primary functions: teaching, research, and community service. The goal in every of those three categories is to encourage or assist learning. Because the ratio of youth is expanding in countries like India, China, Thailand, and other Asian countries, lifelong learning includes young children. Because of the large number of non-literates and neo-literates in these countries, many lifelong learning programs are offered by various agencies, including universities, non-governmental organizations, and government departments. Their focus continues to be on adult literacy, continuing education, and skill training. Learning that is adaptive, diversified, and available at many times and locations is referred to as lifelong learning. It encompasses all present, purposeful learning activities, whether formal, non-formal, or informal, with the goal of acquiring information, skills, or abilities. Womb-to-tomb learning is defined as learning that extends beyond traditional schooling and into maturity. Womb-to-tomb learning has a long and rich tradition in India. Womb-to-tomb education has long been a part of Indian society.

2.4.2 Objectives of Lifelong Learning Education

Lifelong learning encompasses formal, informal, and informal learning, which includes learning at an early age. The following objectives are part of the lifelong learning strategy:

1. to supply everybody with a high-quality course of study and coaching that meets their needs, with a special stress on the requirement for lifelong talent upgrading for those with minimal education;
2. to contribute to India's development as a sophisticated knowledge-based society with property economic development;
3. to foster cooperation and quality between education and coaching systems among Indian states in order to ensure sensible environmental protection for future generations through lifelong learning;

4. to strengthen lifelong learning's contribution to social cohesion, active citizenship, intercultural dialogue, gender equality, and personal fulfillment;
5. to cultivate motivated, engaged learners who are able to face tomorrow's difficulties in addition to today's;
6. to encourage people of all ages, as well as those with special needs and underserved groups, to engage in long-term learning, regardless of their socioeconomic status. Lifelong learning has currently become a distinguished topic within the world discussion of education and coaching.

Different agencies within the country, such as universities, non-governmental organizations, and government departments, provide lifelong learning programs. Their focus remains on adult skills, continued education, and talent coaching, possible because of the large population of illiterates and neo-literates. To integrate formal, non-formal, and informal learning processes, several lifelong learning departments are in operation.

2.4.2.1 Issues and Challenges in Promoting Lifelong Learning Education

A society cannot be deemed an information society unless its voters have access to womb-to-tomb learning facilities and opportunities. There's little question that universities play a crucial role in encouraging womb-to-tomb learning that meets the expectations of adolescents who see education as a force that will facilitate them in reworking their lives. Adult access to teaching, on the other hand, is turning into an increasingly crucial determinant in social group and economic success. Womb-to-tomb learning is well acknowledged and delineated within the Indian academic surroundings as a tenet and overall goal of education.

The core goal of building a literate and learning society has not been met, according to assessment of existing womb-to-tomb learning programs in the Republic of India, because of program constraints and an absence of funding.

There is a scarcity of program innovation, documentation, and distribution. This might flow from to program managers' skilled coaching being inadequate and of poor quality.

The policy should be outlined; the idea should be processed; and therefore the policy should be coupled to the concept of academic policy agenda.

Lifelong learning is presently used as associate umbrella term to cover basic accomplishment, post accomplishment, continued education, and extension programs offered by varied organizations, refresher/continuing courses offered by skilled bodies, non-public establishments, and businesses; but it's not formed as an overarching learning framework. This is largely because of an absence of interconnections between various areas of education, as well as recognition and validation of earlier learning.

2.4.3 PROMOTING LIFELONG LEARNING EDUCATION: ISSUES AND CHALLENGES

Lifelong learning takes a holistic read of education's performance in a person's life cycle. The higher education system can play a major part in this method.

- Lifelong learning departments ought to move from the edge to the foreground and play a key role within the development of human resources, notably in terms of creating a skilled work force within the field of womb-to-tomb learning.

- Adult education ought to continue until 100 percent attainment is achieved. To consolidate the advantages created throughout the lifelong learning, we must always produce neo-literacy and post-literacy programs.
- Currently, educational institutions' roles in supporting learning are limited to what they provide directly to students; nevertheless, they can make a significant contribution by providing early and ongoing teacher coaching, womb-to-tomb learning analysis, and community learning opportunities.
- In order to contribute to pedagogy quality and information development, higher education settings ought to be engineered. A stronger framework and conditions for cooperation between instructional establishments, businesses, and alternative necessary entities ought to be made for this goal.
- Short-term coaching courses, workshops, seminars, summer colleges, and other similar events should be used to advertise acquisition programs for diverse target groups in order to improve their knowledge in their specific specialties.
- Among other things, awareness programs in acquisition, post-literacy, continued education, sanitation, setting, national integration, international brotherhood, gender equality, family life education, nutrition, population education, increasing incomes, and occupation skills for youth seeking employment or self-employment ought to be unionized.
- Opportunities for help and content for adolescents and adults selecting education programs and collaborating in period of time learning should be improved.
- All sorts of education and learning ought to be supported and engineered for a person's information, skills, and competences.
- Policymakers ought to place a high priority on developing skills toward the conclusion of every stage of education. Gender inequities should be minimized at every level. Policy on achievement ought to be aligned with community needs.
- Programs and funding ought to be tailored to the needs of the poorest folks.
- Gender equality ought to be prioritized through teacher education and secure college environments.
- As a part of a womb-to-tomb learning approach, governments ought to dramatically enhance adult learning and education opportunities.
- Occupational learning development ought to incorporate a technique that has community engagement and native cultural life as native learning content.

2.5 CONCLUSION

To improve our economic, political, intellectual, and cultural performance, lifelong learning education is the foremost effective means of building the abilities of openness, flexibility, freelance thinking, creativity, innovation, and leadership. As a result, schools and universities ought to play an important role in preparing student and non-student youth to adapt to changes and gain new skills in response to dynamical social and national demands. Engineers' careers embrace lifelong learning, notably through skilled development needs. However, not all engineering graduates become skilled engineers; several can work within (or outside) the sector while not under official superintendence or assessment of their continuous commitment to and engagement in lifelong learning. Educators' expectations, on the other hand, stay unchanged: we should teach all students to own the skills and attitudes needed to be lifelong learners.

REFERENCES

1. E. Faure, Learning to be: The world of education today and tomorrow, UNESCO, Paris, 1972.
2. J. Delors, Learning: The treasure from within, UNESCO, Paris, 1996.
3. C. Knapper and A. J. Cropley, Lifelong learning in higher education, Kogan Page, Routledge, London, 1985.
4. OECD, Lifelong learning and human capital, Policy Brief, July 2007. Available at http://www.oecd.org/dataoecd/43/50/38982210.pdf, Accessed 12 July 2011.
5. European Union Outcomes for Graduates Dublin Descriptors, A report from a Joint Quality Initiative informal group, 2004, www.jointquality.org, Accessed 12 July 2011.
6. ABET, Inc., www.abet.org, Accessed 12 July 2011.
7. CEDEFOP, The development of national qualifications frameworks in Europe. Luxembourg 2009, www.cedefop.europa.eu/en/files/6104, Accessed 20 Nov 2011.
8. J. C. Dunlap and S. Grabinger, Preparing students for lifelong learning: A review of instructional features and teaching methodologies, Performance Improvement Quarterly, 16(2), 2003, pp. 6–25.
9. R. Koper and C. Tattersall, New directions for lifelong learning using network technologies, British Journal of Educational Technology, 35(6), 2004, pp. 689–700.
10. S. M. Lord, C. Stefanou, M. Prince, J. Chen, and J. D. Stolk, Student lifelong learning outcomes for different learning environments, Proceedings of the 2011 ASEE Annual Conference, Vancouver, British Columbia, Canada, June, 2011.
11. P. Sloep, J. Boon, B. Cornu, M. Klebl, P. Lefrere, A. Naeve, P. Scott, and L. Tinoca, A European research agenda for lifelong learning, International Journal of Technology Enhanced Learning, 3(2), 2011, pp. 204–228.
12. J. Meredith, Launching your Career: Lifelong Learning-your Key to an Enjoyable and Rewarding Career, The Gold Series: Book 4. IEEE–USA E-Books, 2009.

3 Transformational Leadership
The New Mantra for Pedagogy in Higher Education

Padmanabhan Krishnan
Vellore Institute of Technology
Vellore, Tamil Nadu, India

CONTENTS

3.1   INTRODUCTION

The picture of transformational leadership is unclear. To make a blurred picture clearer, with the advent of artificial intelligence, big data analytics and machine learning there is a need to change the disciplines to suit the times of change. Hence, the need for transformational leadership in higher education. In the words of Swami Vivekananda, "The duty of a leader is to make his students think, not just learn. Purpose is what you have when no one is looking at you."

Ayam Nijah paro vaetti, ganana laghu chetasam, udaara charitaanaam tu vasudhaiva kutumbakam.

This one is mine; this one is somebody else's. Thus think narrow-minded people. For broad-minded persons, the entire nation/universe/globe/world is one. Priorities must be broader and not narrow minded. Goals must be specific but not singular.

DOI: 10.1201/9781003083160-4

A broader set of organizational goals to achieve a collective set of objectives leads to more dissemination of knowledge once it is achieved.

3.2   ETHICS FOR PROFESSIONALISM

Ethics is a moral philosophy – a branch of philosophy that involves systematically defending and suggesting concepts of good and bad behaviour in a civilized society. Ethics, along with the nature of aesthetics, addresses issues of a value-based system, and these fields cover the branch of philosophy called axiology.

Ethics solves questions of morality by defining the factors such as good and evil, right and wrong, virtue and vice, justice and crime. As a field of intellectual curiosity, moral philosophy is related to the fields of moral psychology, descriptive ethics and value-based theory.

Three major areas of study within ethics are recognized:

1. Meta-ethics, concerning the theoretical meaning and reference of moral propositions and how their truth values can be determined;
2. Normative ethics, which addresses the practical means of determining a moral course of action, and
3. Applied ethics, which seeks to find out what a person is permitted to do in a specific situation or a particular field of action.

These are some of the ethical basics required for professionalism and any form of leadership that is aspired to.

3.3   VALUE SYSTEMS FOR LEADERSHIP

Becoming a great leader is a long journey and one that should influence our career. The sooner we start focusing on these leadership values, the faster we will become the leader that we want to be:

The 13 core values that a leader must possess are

1. empower and development
2. vision
3. communication
4. reinforcement and influence
5. empathy
6. humility
7. passion and commitment
8. respect
9. patience
10. resilience
11. honesty and transparency
12. accountability
13. integrity

To be a great leader one must have prior experience as a mentor. Mentoring is a process through which an individual offers professional expertise as well as ethical

FIGURE 3.1 The many impactful fields of mentoring.

and moral support to a less-experienced colleague. A mentor serves as a teacher, counsellor and advocate to a protégé. Mentoring results in a mutually beneficial professional relationship over time (Figure 3.1).

What distinguishes a great leader from a poor one? All animals are equal, but some animals are more equal than others – George Orwell said this in his *Animal Farm,* which was a farce on the Russian revolution (Orwell, 2017). Talent, skills, priorities and effective implementation distinguish a great leader from an average one.

In the journey towards achievement of goals, change is inevitable and should be intrinsic first. We must just embrace change and proceed further. Deep rootedness cannot be disrupted as it would pose a threat to sustainability. Short-term goals can do with disruption, but long-term goals depend on sustainability. So the change that we embrace must be manageable. Undesirable changes are to be avoided and ignored effectively, as they would lead to detours and waywardness in the journey to milestones.

Evolution towards artificial intelligence is arduous, and the tortuous path towards AI and the role of a transformational leader in leading a team towards its implementation poses challenges. Trust is the major factor, and it has many facets (Figure 3.2). A transformational leader must have the trust and loyalty of his subordinates to catalyse change, capitalize on strengths and achieve the goals.

The model for a transformational leadership is related to the five wheels of activities. Some of them are a) leveraging everyone's emotional intelligence, b) having short-term and long-term goals and plans, c) being able to strategize and achieve in multiple organizational goals and d) being able to perceive change, strategize and achieve (Figure 3.3).

A distinction has to be made between transactional and transformational leadership here. The differences between transactional and transformational leadership are enumerated here.

FIGURE 3.2   The core necessities and relationship of various factors to trust.

Transactional leadership is a type of leadership whereby rewards and punishment are used as a baseline for getting the followers immersed. Transformational leadership is a leadership style in which the leader uses his persona and enthusiasm to influence his followers, who develop a passion to do things.

In transactional leadership, the leader lays stress on the hierarchical relationship with followers. But in transformational leadership they lay stress on the values, beliefs and the needs of their subordinates.

Transactional leadership is considered reactive, whereas transformational leadership is proactive.

Transactional leadership works in a well-settled static environment, but transformation is good for a volatile and turbulent environment.

FIGURE 3.3   The five wheels of transformational leadership.

Transactional leadership works for improving the prevailing conditions of the organisation. On the other hand, transformational leadership works for changing the present conditions of the organisation towards a tangible future.

Transactional leadership is bureaucratic, whereas transformational leadership relies on a feel-good factor.

In transactional leadership, there is only one possible leader in a group. This is in contrast to transformational leadership, in which there can be more than one leader in a group.

Transactional leadership is focused towards strategic planning and implementation compared to transformational leadership, which promotes innovation, ideation and implementation.

3.4 VIRTUES THAT HELP IN TRANSFORMATIONAL LEADERSHIP

Transformational leadership needs more virtues than any other form of leadership. Virtues that help in transformational leadership are

- Subject knowledge, communication skills.
- Clearly stated objectives, methods, results and summary.
- Loud and clear dynamic model of creating enthusiasm.
- Linking the subject with a colleague's prior experience.
- Ensuring interaction of the colleagues without distraction (Mission 10X by WIPRO, 2010)
- Avoiding visible mannerisms.
- Based on assessment and feedback, improving story-telling techniques to suit slow learners.
- Maintaining cordial relationships to get your story across.
- Being an inspiration and impressing the subordinates
- Being innovative and unbiased.
- Being a lifelong learner.
- Not pushing your contributions and side-lining the pioneering references.
- Putting your work in a realistic context and embedding it suitably to showcase your contributions to enable collaboration.

Remember, a mistake that makes you humble is better than an achievement that makes you arrogant.

3.5 STRATEGIES FOR ENHANCING STAFF AND STUDENT MOTIVATION

Challenge yourself before challenging any one to follow your challenge.

- Build on strengths first, and then address weakness.
- Offer choices; prepare a roadmap that is subject to review down the line as change is inevitable.
- Provide a secure environment.

FIGURE 3.4   Cart before the horse.

- Teach them how to make their tasks manageable.
- Even at peak responsible activities the skills must provide a comfort factor for easy execution.
- Use rewards and punishments with caution.
- Help colleagues develop an internal locus of control and avoid power struggles.
- Be clear and not ambiguous.
- Develop creativity through activities.
- Teach them introspection and self-evaluation.
- Seek their attention and make them competitive and more perfect. (Mission 10X by Wipro, 2010)

Finally, are leadership and teaching instructive? Didactic? Directive? No. Leadership is a developmental activity. In a developmental activity don't put the cart before the horse (Figure 3.4), or else it could be a tragedy like the *Titanic*, a ship that was considered unsinkable though lacking in reliability tests that should have preceded the inaugural – but terminal – journey.

3.6   TEACHING/LEARNING EXCELLENCE THROUGH TRANSFORMATION

The rubrics of the new education policy by the Indian government, (New National Education Policy) and the three developmental skills required for a leader are illustrated in Figures 3.5 and 3.6. The new education policy focuses on innovation,

New Education Policy

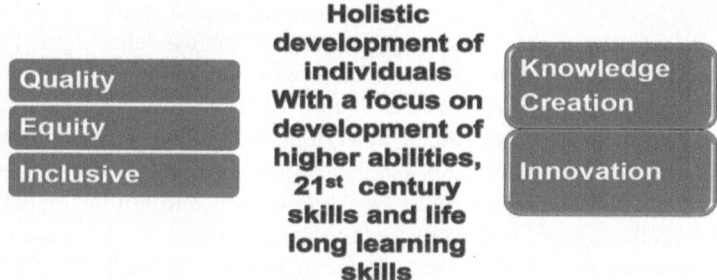

FIGURE 3.5   The rubrics of the new education policy by the Indian government.

knowledge creation and inclusivity in growth. The Government of India has devised the new education policy with inclusive economic and technological growth in mind. The three skills illustrated in Figure 3.6 that every student, teacher and prospective leader must possess can also be expressed as the thought, emotional skills and physical skills that impart a complete training to any aspiring individual. The physical trainings, however, are manipulative. Thought leadership is an outcome of the first two skills. A transformational leader must be a combination of all three. A chain of National Initiative for Technical Teacher Training and Research (NITTTR) colleges in India train the future leaders for mentorship and peering.

The transformation from the old and new Bloom's taxonomy that teaches many levels of intelligence puts creation and innovation content above application and knowledge. Applications are a derivative of creation, synthesis, evaluation and assessment. For a transformative leadership programme, Dave's psychomotor taxonomy, Gagne's

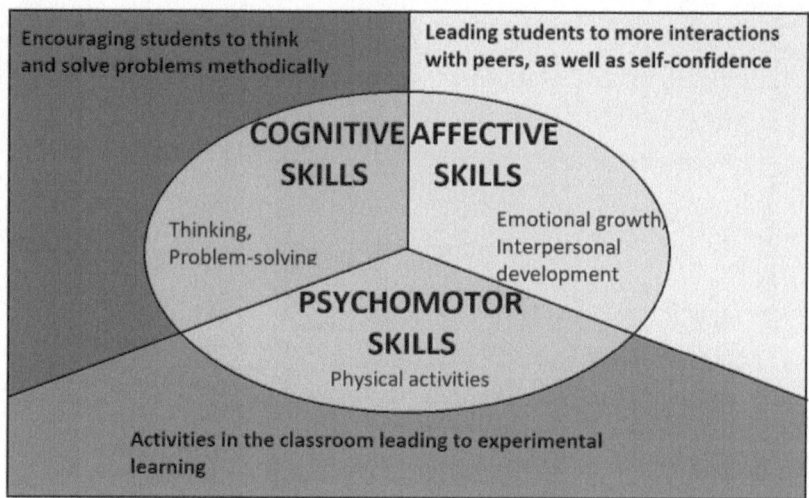

FIGURE 3.6   The three developmental skills required for a future leader.

nine events of instruction for a teacher and the Bloom's taxonomy are included as a mandatory syllabus to make the candidate understand the importance of creation, synthesis, and events of instruction that play a vital role in transformative leadership. Excellence in teaching and leadership evolve from diffuse learning to conceptual learning and implementation. A need for transformational teaching is very desirable in the present scenario. As a developing nation we must add the Guru-Sishya Parampara learning system and adapt western models to suit Indian spirituality and national priorities. Any unskilled or parasitic adaptation will be ruinous. Value and satisfaction, not just commerce, must be the priorities. Success lies in satisfaction; corporate benefits are only materialistic in their outcomes.

The Cambridge model of student outcomes and the attributes that they develop in their graduates are discussed here (LeAP Module, 2019–20). The outcomes and attributes that every student must develop at the end of their studentship are a) analytical abilities, b) collaborative spirit, c) communicative capabilities, d) contextual awareness, e) critical thinking, f) independence, g) being grounded and practical, h) inclusivity and culturally agility and i) interdisciplinary approach and responsibility.

Finally, You are rated by what you produce, not by what you attempt

Skill development for employability is a "much easier said than done" task than an industry gearing up to match an academia's excellence, Figure 3.7. Proper use of education and developed skills in career development lead to the desired transformation, given that the students first become employable. More opportunities are needed for transforming careers and leadership roles.

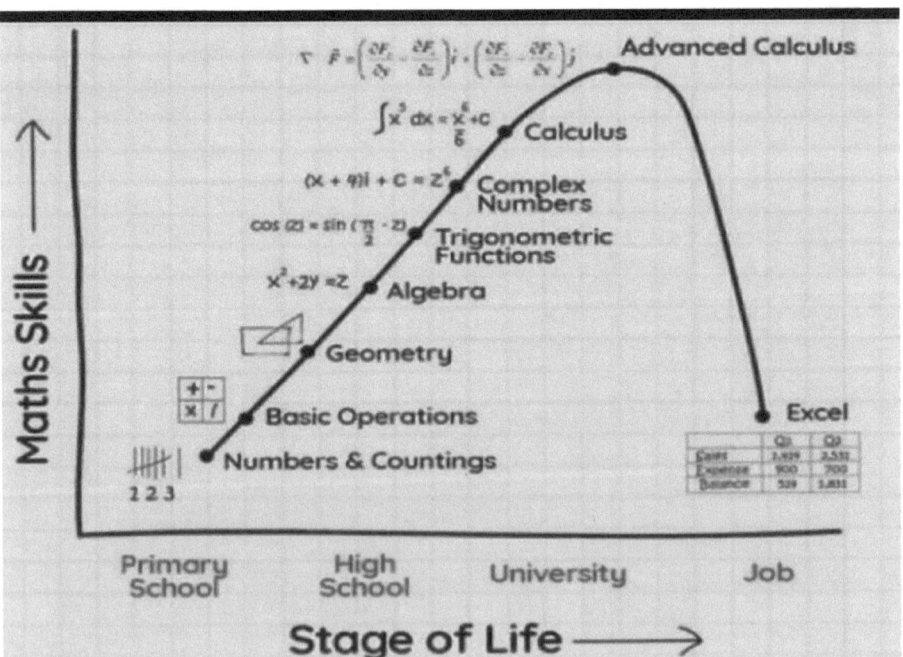

FIGURE 3.7 The main deterrent to transformation in the job market is lack of skill implementation.

3.7 DIGITAL PEDAGOGY

Most of the classroom learning basics are applied to digital learning, expecting the same outcomes or better, though there is much to be done to achieve the desired outcomes.

We have to assess whether the outreach is the same as in classroom learning, which is very difficult to manage. Though Covid-19 has accelerated digital learning, it should not be the story of liberty ships built during WW II that couldn't last the war! Digital learning must be strongly related to sustainability and sustainable goals. Sustainability and professional ethics must be taught at the beginning of every programme so that the long-term goals and ethical necessities required to reach these summits are realised with adequate levels of transformation. Only transformational leaders can achieve these. So, the Covid effect must be understood for sustainable outcomes from not-so-long-lasting blisters to an everlasting bliss, called attainment.

A paradigm transition to undergraduate education can be made using the left-brain thinking and right-brain thinking–based pedagogies. Supervised learning must pave the way to skill-based learning in which learning to think independently is achieved. A revisit to admissions policies and sensitizing stakeholders to impart a sense of professional ethics, sustainability and professional learning is the need of the hour. Learning enablers must be activated, such as heuristic learning, original thought and expression tests, self-learning tools, and experiential sharing and responsibility endurances.

3.8 RESEARCH EXCELLENCE AND TRANSFORMATION

The present structure of education institutions is a socio-cultural hub with faculty, students, technicians and the management. The best of infrastructure, education, innovation and fund generation lead to socio-economic development of the nation. To achieve academic excellence and perform the mandate effectively, research excellence is a must. It is defined by quality publications, patents, funded projects, consultancy projects, tangible products and efficient management. It is important leverage for transformational leadership, which must adopt a rather unconventional approach to achieve excellence. To assess, evaluate and understand the strengths, weakness, opportunities and challenges, one must perform a SWOC analysis and not a SWOT analysis, as challenges are more important than perceived threats.

Research strategies to achieve excellence can be listed based on priorities. Recruitment of competent faculty, creation of interdisciplinary research groups, recruitment of industry faculty and adjunct faculty, recruitment of tenured track for international faculty, creation of research councils, identification of niche research areas, provision for a faculty-friendly intellectual property policy, creation of incubation facility, collaborations with industries, development of industry-funded research labs, collaboration with international industries and academia, attracting international funding for faculty and student exchange and continuous assessment of key performance outcomes are some of the priorities in the guidelines for transformational leadership to achieve research excellence.

The challenges that every transformational leader faces are a) retention of faculty, which is difficult due to openings in different institutions, b) industry faculty recruitment through existing recruitment rules, which is difficult, c) fulfilling the aspirations for foreign faculty, d) aligning faculty with industry collaboration and, e) making every faculty focused on research and being self driven.

In achieving research excellence, roadmaps should be subject to revision and review every six months or a year. All roadmaps should be subject to progressive change every six months. Perceived and disastrous changes must be combatted to avoid skewed exits that prevent us from achieving prioritized and long-term goals. As change is inevitable, roadmaps cannot be milestones from the past.

3.9 ACHIEVING AND SUSTAINING RESEARCH EXCELLENCE

Excellence in research for societal impact can be felt with research that makes a difference.

Addressing problems of current and future societal needs, choosing appropriate targets, fundamental research leading to new discoveries that may take long time but be impactful, research translating into deliverables in reasonable time, trust-building research that lays a strong foundation, research demonstrating capabilities and values – all these lead to excellence.

The reputation of the institution as the preferred destination for faculty, students, industry and other stakeholders and as the knowledge hub for high-quality deliverables that enable excellent minds to work with target oriented objectives and also generate excellent trustworthy manpower to make a difference – all these emphasise the importance of research excellence that requires transformational leadership. These must be the essential mandates of an academic institution. How is this achieved?

- Through identification and motivation of people
- Through a research culture and bringing all on board
- Believing that people make a difference
- Strategic research planning
- Supporting fundamental and applied research
- Identification of target areas
- Providing full support
- Providing interdisciplinary and transdisciplinary communities
- Financial planning with seed grants, research grants (government, alumni, endowment and industry)
- Core research grants
- Industry-relevant research and development
- Start-up research grants
- Intensification of research in high priority areas funds for Improvement in science and technology infrastructure
- Creation of institute/central facilities
- Creation of research/science parks, technology incubators, venture initiatives, consultancy centres and test facilities (SATHI)
- Promotion of impactful consultancy and industry connections
- Research income from societal impact, and
- Commercialization after realization

Vision, people, plan, politics (VP3) should be the motto of every research excellence initiative. It is to be emphasised here that strengthened PhD programmes and strong postdoctoral programmes are the backbone of research. Library resources must keep abreast of latest developments

Some of the additional inputs that aid in the attainment of research excellence are

- rigorous periodical peer review and metrics for evaluation
- linkage to promotion; excellence in research recognition (students and faculty)
- performance incentives and awards
- keeping up pressure to perform
- wider stakeholders and consultation
- moving ahead with demands of time and societal needs
- international collaborations, MoUs, exchange programs, campus drives
- foreign import substitution: through economical indigenization and promotion of local talent

3.10   EDUCATIONAL MODELS AND TRANSFORMATIVE INITIATIVES

Cambridge University's 'go to them strategy' is seen as a very productive method as against other's 'come to me strategy' (https://www.ifm.eng.cam.ac.uk). Some of the educational models for societal and economic development are

- Institutional models: Private held by an individual, family, group or public university
- Organizational model: Held by an individual, family, group, trust or public firm with investments in industries, educational institutions, media and leisure.
- Corporate model: The above with a corporate-management–based approach and outcome-based economic benefits.
- Enterprise model: A collaborative and joint effort between the educational institutions and the students, investors, stakeholders, such as Industries, start-ups, innovators and research laboratories through partnerships (IIC) (Cambridge Business Model Innovation Research Group)

A simple model is presented here for teaching, learning and research. Developmental teaching and learning must be practised as opposed to instructional and directive learning. Learning must be student centric as they are the majority beneficiaries. Applied research is to be carried out for a) slow, academic and sustainable development and b) fast-track research with corporate gains (Industry 4.0 ready) as the two different pathways.

UN sustainability goals and the corresponding web site provide enough resources for the role of transformational leadership in leading the way. (www.sustainabledevelopment.un.org)

Last and highest, leadership must cater to accreditation and ranking by National Accreditation and Assessment Council (NAAC), Accreditation Board for Engineering and Technology (ABET) and Institution of Engineering and Technology (IET), with more in the offing! Recording and documenting your approach, methodology, tools employed, improvements observed, achievements made and feedback given are to be practised to the hilt by any transformational leader. Ranking frameworks like National Institutional Ranking and Framework (NIRF), Quacquarelli Symonds (QS), Times Higher Education (THE) and the like have to accredit and rank your institution based on the rubrics given so far. Proven storytelling success records can be taken as benchmarks for ranking exercises. Repeatability and reproducibility should be the mantra for any consistent and sustainable path and exercise.

An institution of many activities would gain a lot from a centralised Enterprise Resource Planning (ERP) module for all its billing, logging, sharing, document filing, financial record keeping and purchase. A very useful software similar to the details provided in Figure 3.8 has been developed by Indian Institute of Technology, Kharagpur (IITKGP) (https://www.ndl.iitkgp.ac.in). Such a digital portal improves an institution's function and documenting efficacy, which in turn improves its score in accreditation and ranking exercises. It has to be implemented by every institution aspiring for high ranking as it would make the filing process a one-time exercise and avoid duplication of data and details. Digital empowerment is one of the prime duties of any transformational leader.

The Vellore Institute of Technology, which is the number one private engineering institution in India is an Institution of Eminence, having been accorded the status by Ministry of Human Resource and Development (MHRD), India (www.vit.ac.in). The leadership here, headed by Honourable Dr. G. Viswanathan, is highly transformational,

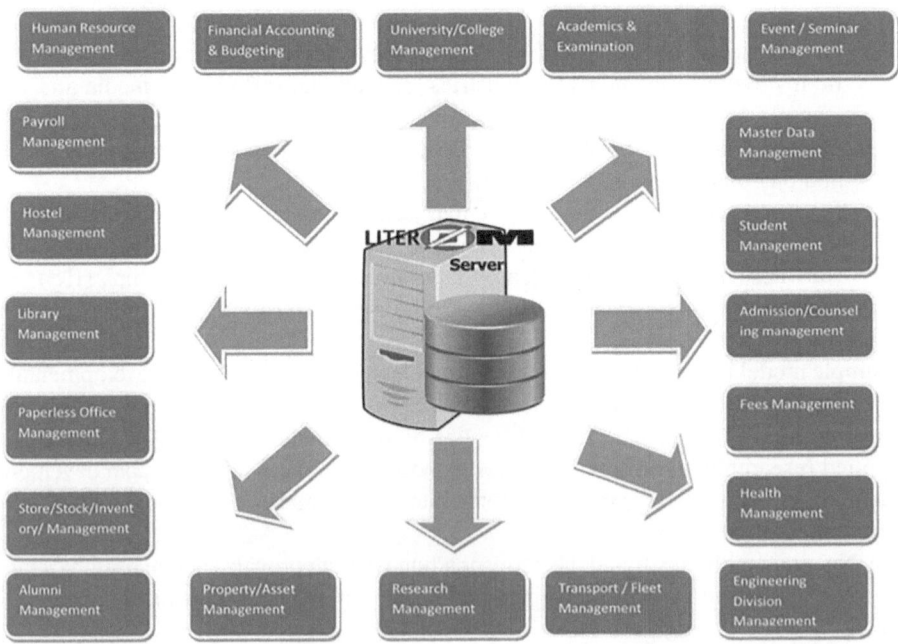

FIGURE 3.8   A centralized ERP system for effective transformational leadership.

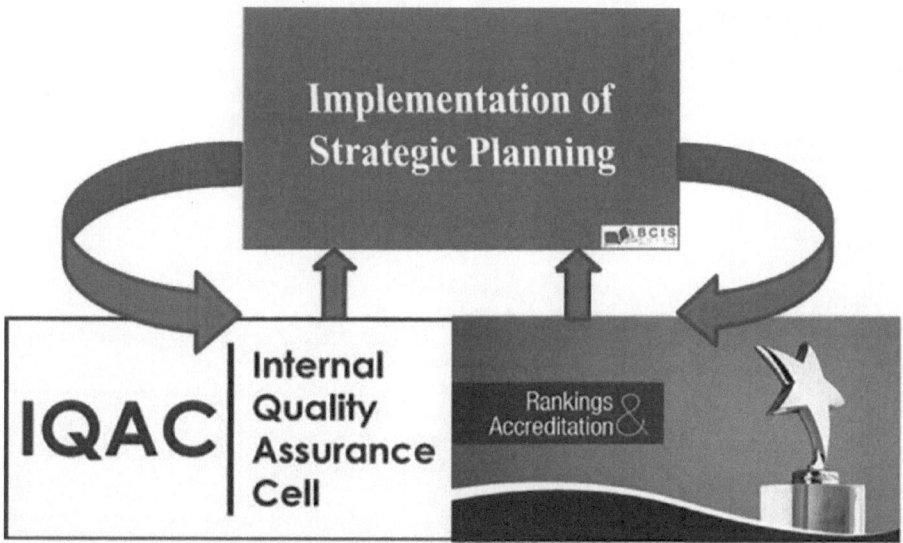

FIGURE 3.9   The quality circle between planning, quality, implementation, and accreditation.

and the rigorous proactive activities of strategic planning and implementation coupled to the Ranking and Accreditation Council (RAAC) and the Internal Quality Assessment Cell (IQAC) can be termed as one of the best in the country. A schematic of the practised model is shown in Figure 3.9, which involves good feedback to the RAAC and the planning team from the IQAC.

Finally, a good and short motto for every aspiring and successful institute of engineering would be vision, value and visibility – the three Vs for inclusive growth.

Aham brahmasmi and Jai hind!

ACKNOWLEDGEMENTS

I thank my colleagues at the IITKGP/MHRD/LeAP programme, University of Cambridge for the association and my wonderful management at Vellore Institute of Technology for all the experience, support and encouragement.

REFERENCES

1. George Orwell, Animal Farm, Finger Printing Publishing, India, 2017.
2. Mission 10X Training Module, by WIPRO, India, 2010.
3. The New National Education Policy, https://www.education.gov.in
4. Leadership for Academicians Programme, LeAP, IITKGP/MHRD/IfM, University of Cambridge, Training Module, November 2019– January, 2020.
5. Cambridge Business Model Innovation Research Group (CBiG), https://www.ifm.eng.cam.ac.uk
6. www.sustainabledevelopment.un.org
7. National Digital Library of India, https://www.ndl.iitkgp.ac.in
8. Institutions of Eminence, www.vit.ac.in

Section II

Ethics

4 Ethical Perspective on Engineering Education
A Systematic Education Approach

Shivangi Sahay
Indira Gandhi Delhi Technical University for Women
New Delhi, India

Anju Bharti
Maharaja Agrasen Institute of Technology
Delhi, India

CONTENTS

DOI: 10.1201/9781003083160-6

4.1   INTRODUCTION

4.1.1   ETHICS

Ethics is system about moral principles. Ethics states what's good for people and society. *Ethos* is Greek word that means character, custom, habit, or disposition (Chhabra, 2016). Ethics may be a philosophy that involves protective, organizing, and recommending concepts of right and wrong behavior. These are some of the rules or the principles that are considered either right or wrong and are imposed by society. There are different theories of ethics: meta ethics, normative ethics, and applied ethics (Anscombe, 1958). Meta ethics tells about existence and meanings of ethics.

Ethics deals with fundamental problems with the practical cognitive process and its major concerns by which human actions will be judged right or wrong. Ethics, a product of society, takes the thoughts and concepts of the many people to form an ideology. Many ethical issues reflect the connection between the individual and the group.

4.1.2   ETHICS IN ENGINEERING EDUCATION

Ethics depend upon the ethical obligation that a person may feel. The study related to ethical ideals, character, regulations, and relationships of people and organizations concerned in technological activity may be termed *engineering ethics*. An engineer who works for an organization, will encounter a few moral problems through conceptualization of a product, problems coming up in designing and testing. Troubles may also come up in manufacturing, sales, and services. The moral decisions and moral values of an engineer are taken into account because their decisions will impact the quality of products and services. The values of engineer are judged by the goods manufactured, trust level of shareholders in the corporate, process benefiting the people, regulations in the profession and industry, the prevention of pollution, etc. Engineers, like others, have to observe a set of ethics about the way to live without getting morally degraded (Chhabra, 2016). The curricula must involve concepts with the ethical aspects such as

- to respect everyone, including self
- to respect the rights of others
- to keep the promises made
- to avoid causing unnecessary trouble to others
- to avoid from dishonesty.
- to show gratitude toward others and inspire them

Morality advises respect for all and entails being honest and not causing harm by dishonesty to humanity. A well-organized mechanism of engineering helps speed up the work with the assistance of technology optimizing resources. Ethics binds society with its principles, which are associated with the moral standards of people at large. Ethics helps an engineer with their extraordinary performance suitable for the society.

The implementation of engineering ethics in engineering is vital for society's welfare. It is the study of choices, policies, and values that are required part of engineering practice and research. It's the system of moral standards that examines the obligations by engineers towards society and customers and consequently the career itself.

Engineers are expected to exercise ethical characteristics such as honesty, impartiality, fairness, equity, etc. Their devotion to the safety and welfare of the society must be reflected. Engineers must adhere to ethical conduct as a part of professional behavior in their work performance. Ethical conduct in engineering is a very important to the profession. The ethics of a profession are all about the foundations or standards governing the conduct of the members of the profession.

4.1.3 ETHICAL CONDUCT OF ENGINEERS

Engineering is a learned profession, and maintaining standards of honesty and integrity reflects dedication to the work. Engineering, as a system, impacts the quality of life of people (Kumar, 2015). People rely on engineers to provide them with safe and reliable goods and services. Their performance highlights a standard principle of ethical conduct (Kumar, 2015). The mistakes of unethical and incompetent engineers could cost many lives. Engineering ethics will be able to prevent such grave consequences and will help give meaning to the real efforts of engineers. The respect and reputation of the engineering profession requires them to be ethical and responsible. Engineers fulfilling their professional duties (National Society of Professional Engineers, 2019), should also be able to

1. Take care of welfare of the public, which includes safety and health.
2. Perform services in areas they are most competent.
3. Issue statements for public in an objective and truthful manner.
4. Be faithful agents for employers and clients to win trust.
5. Avoid deceptive acts.

Engineers have certain obligations toward their profession, their customers, and society. Following is the code of conduct of ethics that should be maintained and practiced:

- Integrity.
- Objectivity toward life.
- Professional competence.
- Confidentiality in work for organizations.
- Balance in professional behavior.

Ethics is fundamental in problem solving in engineering. The principle of engineering codes of ethics is for protection and the prevention of all types of harms to humanity. An engineering course must contain these features specifically. Engineering ethics may be categorized as individual, professional, and social.

Engineering ethics has two parts: micro ethics and macro ethics. The relation between individuals and the engineering profession is highlighted by micro ethics. The social responsibility of the profession and societal decisions about technology is collectively highlighted by macro ethics (Indian Engineering Services General Studies, 2017).

4.1.3.1 Micro Ethics

Micro ethics has two levels, individual and professional, as seen in Figure 4.1. Individual ethics are about equality, honesty, and integrity, etc., whereas safety and quality, etc. related to product manufacturing as a final outcome are part of professional ethics.

4.1.3.2 Personal Ethics

Personal ethics judges qualities like truthfulness, decency, honesty, technical qualities and responsibility. It also consists of professional value and personal qualities which are as follows:

4.1.3.3 Technical Ethics

Engineers make the technical decisions and judgments at a micro level analysis of individual technologies or practitioners.

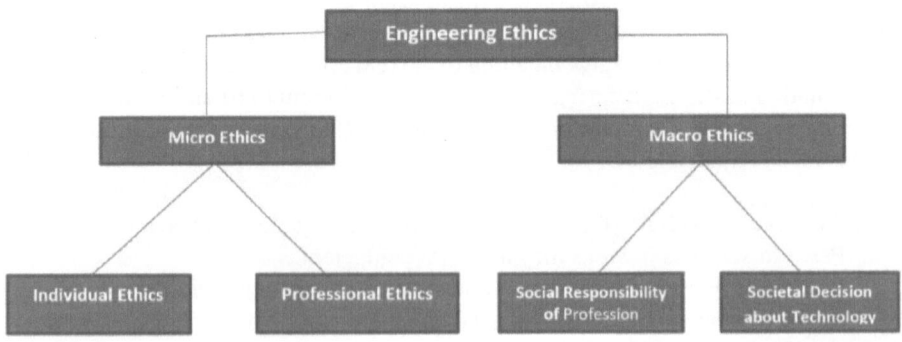

FIGURE 4.1 Engineering ethics.

4.1.3.4 Ethical Responsibility

Ethical responsibility helps in making wise choices. There's a willingness to have interaction with others regarding the crucial choices.

4.1.3.5 Professional Ethics

The knowledge possessed in this profession creates the sense of moral responsibility. There is a responsibility toward maintaining professional relationships between engineers and other employees.

4.1.4 ENGINEERING DESIGN ETHICS

Issues of engineering design ethics arise while designing technological products, processes, systems, and services (Pahl and Beitz, 1996). Engineering design concerns are about user autonomy, safety and sustainability, and privacy. Ethical concerns in technology are focused on the needs of the user. Technology helps in design, and it underlies many ethical issues and challenges to get the product ready for operation.

Design choices are manifested in different stages in the process, and they influence all ethical issues in the technology (Stitt 1999). Engineers must obey the engineering codes of ethics that can fulfill the requirements of design.

4.1.5 ETHICAL PERSPECTIVE TOWARD SAFETY IN ENGINEERING EDUCATION

Engineers are expected to maintain not only the values of the safety of the general public but also of health and overall welfare.

Following are the fundamental values:

4.1.5.1 Safety

The responsibility of this profession is to provide safety and protect people from harmful effects of product design. Safety of people is of utmost importance as per the professional ethical engineering societies. Safety means that any product or facility should avoid injury, damage, or risk to society. Risk is the chance of harm or loss to people.

4.1.5.2 Case Study

The Bhopal gas leak disaster is one of the world's worst tragedies and industrial disasters in history. It showed irresponsibility toward the safety of people and environment. There was the absence of risk assessment and ethical behavior within the UCIL plant.

4.1.5.3 Public Welfare

Public welfare is also about meeting the standards for safety and health at work. It is the responsibility of engineers to ensure quality services and products associated with work projects. It can be done through identifying various ethical, economic, cultural, legal, and environmental issues.

4.1.5.4 Health

The health of the public may be harmed by toxic elements, pollution, defective design, etc. If certain standards are maintained as a protocol, harmful emissions and usage of toxic material can be minimized. Any potential risks need to be discussed with the authorities.

4.1.5.5   Macro Ethics

Macro ethics comprises of professional and social ethics. Social justice, sustainability, poverty, and bioethics are the concerns to be addressed by the engineering profession. Following are parts of macro ethics:

4.1.5.5.1   Professional Ethics

Engineering professionalism and social responsibility are the components of professional ethics. Collective action will also be applicable internationally.

4.1.5.5.2   Social Ethics

Here, the issues related to technology which benefits the society are included. Decisions are made considering the ethical implications of sustainable development, public policy, healthcare, and data communication technology. issues related to robotics and autonomous systems are also included in social ethics.

4.1.5.6   Role of Professional Bodies

The role of professional bodies is to develop codes of ethics. Such societies enables individuals to connect to professional and social ethics. Professional societies' leaders can influence individuals to adapt to engineering culture.

4.2   NEED TO CHANGE EDUCATIONAL APPROACH

There is a need to modify engineering curricula as a response the advancement of technology in every field. Also, there's a need to promote motivational changes in students' attitudes and to enhance their performance. These phenomena are associated with societal changes, along with the proliferation of computers and data technology. The curricula should involve more hands-on learning, laboratories for practical work, textbooks that contain realistic examples, computer-based learning, experimental and computational techniques, and live case study. Better ways of communications must be inculcated in courses that address frame of mind.

As the technology is advancing, academic transformation is the need to make teaching and learning more advanced at the professional level for graduate engineers. Traditionally, the transmission and acquisition of data through education were based only on the instructive approach of teaching and learning (Standford and Keating, 1998). The conventional educational or tutorial model and university research project for the professional requirements has evolved follows (Stanford and Keating, 2004):

Curiosity in mind ends up in → basic research or to explore → knowledge gained → teaching → learning → application in practice.

As it's a profession that solves societal problems, engineers participate in this endeavor as creative problem solvers (Baruh, 2012). So the main focus of students and educators shouldn't be narrow and limited to only specific course content, knowledge, and the tools for simply earning a livelihood. Rather it should be ready to play a much bigger role within the context of a wider audience, such as societal values, the role played by educational institutions, professional or organizational approaches to problem solving, and the political climate and system under which the engineering profession operates (Rajan, 2017).

Educational institutions and academicians should develop curriculum elements that are apart from the conventional science-driven model of technology. Education should assist students in developing competencies in areas like teamwork, communication, lifelong learning, innovation and introduction of newer technology for overall development of the society (Cruz et al., 2019). The effectiveness and competence of curriculum elements must be evaluated from time to time (Bush, 1945).

Innovations are going to be essential to achieve qualitative changes in education, as in other sectors. The changes are desirable for enhancing efficiency and improving the standard of learning opportunities (Stanford and Keating, 2004). Education brings innovation in society by developing and encouraging the proper skills. Talents that may be fostered through appropriate teaching may include creativity and imagination, critical thinking, and to adopt practices like entrepreneurship education (Organisation for Economic Co-operation and Development, 2016). Improvement in education bring change in technology, and its effect is shown in quality and productivity. Engineering education needs improvement to provide work efficiency, increase quality, and foster equity in the way other sectors have done to bring improvements. Governments should influence smart strategical policy for education and create an innovation-friendly culture to enhance efficiency.

4.2.1 Innovation in Education: Innovation in Economies and Societies

Innovation can bring change into stagnant markets and make organizations capable of accepting change in external environments (Damanpour and Gopalakrishnan, 1998; Hargadon and Sutton, 2000). Innovations is the prime factor in developing competitiveness in a globalized economy. It compels students to learn and think to solve many problems. It further enhances the thinking capacity to do the things in different way. This is required in the business world. Innovative policies and theories have focused on the business sector mainly (Lekhi, 2007). Organizations bring innovation to remain ahead of competitors in business. They can launch new products/services on digital platforms, improve efficiency, and pursue uniqueness in marketing activities to guarantee survival in the competitive scenario. Innovation in engineering education may earn significant technological advancements resulting in social welfare gains. Educational innovation improves the education establishment and learning outcomes (Organisation for Economic Co-operation and Development, 2016). Education is supposed to enhance equity and equality in learning outcomes and is relevant to society and national economy.

4.2.2 The Instructive Approach to Teaching and Learning

The conventional approach of education was only to ensure understanding of the reading material presented to students. There was hardly any scope of improvement in educational approach. It was only to transfer knowledge to the students. For better understanding of knowledge transmission to the students, supplementary learning activities were reinforced, such as working in laboratories, to gain experience while working in the field (Knowles, 1973). Previously, the instructive approach of engineering education to teaching and learning was followed. Knowles (1973) has described it as a conventional and instructional process of education. It is basically a way of transmitting knowledge from teacher to student.

4.2.3 NEEDS-DRIVEN MODEL FOR TECHNOLOGICAL INNOVATION IN ENGINEERING EDUCATION

Engineering curricula should be based on scientific research for effective technologies for the purpose of industry and society as a whole. The needs-driven and creative engineering method is the foremost requirement for responsible leadership for the sake of innovation and technology development.

Curricula and design projects developed for the students must have authenticity so students can develop informed design practices and become reflective decision-makers. Teacher education has to address teacher beliefs together with pedagogical content knowledge. To facilitate three-dimensional learning, teachers must be supported in developing competencies as coaches within the classroom with abilities to elicit, recognize, and answer student thinking. They're also expected to take care of students' social and emotional needs (Purzer and Quintana-Cifuentes, 2019). There should be awareness regarding the gap between academic research and actual practice of professional engineering for needs-motivated technology development. Concepts and theory must be included in curricula for better understanding and to search out the solutions of the problems.

4.2.4 NEED FOR ADVANCED PROFESSIONAL EDUCATION FOR ENGINEERS IN INDUSTRY

The differences between two approaches were noticed, one in scientific education for research and another in imparting professional education for creative engineering practices. Hollister (1949) described the application of creative engineering methods in which these concepts are used for educational development in addition to meet these real-world needs. Implementation of creative solutions in engineering is needed for engineers to prove themselves responsible professionals.

Advanced professional education relies on a content–process model to unravel real-world societal needs that were not recognized previously. This would a part of educational development of competent practitioners (Standford and Keating, 1998). Advanced professional education develops technical competence, intrinsic creativeness, and leadership potential within the candidates.

The curricula for graduate engineers should change with time as a number of the traditional principles of teaching and supervision do not match with the developmental growth and needs in the industry, especially for engineers who want to pursue professional leadership careers.

A systematic approach of learning will help competent graduate engineers to practice their knowledge and skills in several ways. It would certainly help them in finding in work, finding solutions to problems, program making, policy and strategy making, and strengthening leadership qualities with professional responsibility. Those are the professional dimensions of systematic engineering practice and professional leadership. Innovations in a systematic approach in education include the following for engineers:

- To have technical competence
- To be creative in solving problem and systems thinking in an innovative way
- To maintain responsibility professionally
- To be creative and utilize needs-driven collaboration and to be confident for professional leadership of multidisciplinary fields
- To be ready for problem-finding and visualization (needs-finding)
- To have capability of program making and strategic thinking
- To solve problems in ethical ways, which is safer for society, and to help in policy making and value judgement, along with economic issues associated with technology.

4.2.5 Developing Students' Aptitude through University and Industry Collaboration for Practical Usage of Concept

The curricula for undergraduate engineering students should be constructed in a way that their aptitude/skills must be enhanced. It can be in the form of creativity in activity and generating new opinions, team coordination, problem solving, and developing leadership qualities. Apart from conventional knowledge, these skills need to be inculcated so that if they get an opportunity, they will be able to give better performance in every field.

The universities must take the responsibility for using innovative and suitable approaches to serve their students with in-depth technical knowledge with professional skills. In this manner the education of engineers will be fulfilling and satisfy the needs (Aizpun et al., 2015).

Today's trend is to keep university research linked with industrial innovation for a positive purpose. It has been widely recognized as useful for knowledge sharing among academics and industrial practitioners. The transfer of academic knowledge into the industrial domain is a remarkable step taken by the academicians and educators. It is really a surprising fact that almost everyone neglected the role of collaboration between the research-based university educations with industry. The main role of universities is to provide education and the creation of knowledge but collaboration with industry will boost the overall development and improve competence among engineering students (Arvanitis et al., 2008). The industrial firm and university research group may collaborate for educational involvement and learning practically. This will enable relational learning and innovation development in university and industry relationships. The curricula should include the scope of collaboration for educational involvements in the form of projects, jointly organized courses, and tailored degree courses. Studies have revealed that educational collaboration is facilitating new knowledge and developing university and industry relationships. Thus, educational collaboration may consist of providing educational activities and joint training to students, or it may work on different projects (Arvanitis et al., 2008). This relationship between academia and industry will help in overall development of students, institution, and country as well.

Transferring knowledge may be inter-organizational which is the primary driver of innovation (Tsai, 2001), where both partners share their knowledge and information. It may be inherent or experimental in nature. The sharing of information requires a trusted atmosphere, as it has economic value and potential competitive advantage as well (Santoro and Saparito, 2003).

4.2.6 LEARNING TO ADAPT CULTURE OF ORGANIZATION: EDUCATION FOR INNOVATION AND THE LEADERSHIP

The curricula should be made to reflect the combined effects of advanced technology education with a professional's experiential growth in industry. Students should be made to learn about adapting the culture of the organization through total involvement in its system and prepare themselves as leaders. The aim is to provide an integrated practice-oriented advanced professional program to develop competent qualities. They should have innovative leadership capabilities with creative engineering work to satisfy societal/industrial needs.

The curricula should also have the key features associated with learning organizational culture that can be utilized to help in the growth of the company. These key features include continuous learning, knowledge generation and sharing, systems thinking, learning culture, workplace flexibility, and valuing employees (Dejmal, 2020). The overall approach of education of teaching and learning should provide a new direction toward developing the capability of engineers so that they establish themselves as technically competent innovators and leaders in industry.

4.2.7 EMBEDDING ENTREPRENEURSHIP INTO THE CURRICULUM (OECD, 2016)

It is considered very motivational if students get to experience working at an enterprise as interns while studying at university. It will enhance employability and entrepreneurship skills. The concept of entrepreneurship will suggest a career option as a business startup or self-employment. The educators will get a chance to assist students find out about enterprise and entrepreneurship. They are able to easily relate concept usage to the sensible world (Kelly, 2011). There's a lot of scope for students to study subjects related to business activity such as tendering exercises or sharing new ideas. The students benefit by gathering knowledge regarding business issues and links with the profession. They are obsessed with formal and informal evaluation of the task given to them (Kelly, 2011).

The inclusion of entrepreneurial concepts in curricula will generate career options and will provide opportunity and confidence to budding engineers to develop their own business. The ideas of research students may be converted into technological enterprises that may stand apart from competitors. Business skills will enable them to become self-employed or work on a consultancy basis. Hegarty (2006) argues that "entrepreneurship is learned phenomena." Universities encourage opportunities for business to flourish (Gibb, 2005; National Council for Graduate Entrepreneurship, 2008). It will further strengthen practical skills such as oral and written communication and personal effectiveness skills. These skills are important to organize oneself for a rapidly changing economy and to manage workplace uncertainty. The students will learn flexible working patterns for challenging careers.

4.2.8   SYLLABUS TO ASSIST IN ACCREDITATION REQUIREMENTS

Engineering education curricula must keep the pace with the world where knowledge is spreading rapidly. Because the time and need keeps on changing, curriculum development and updating should be a never-ending process (Stabback, 2016). Educators have recognized the existence of competition internationally in teaching and learning as teaching and learning are complex activities evolving from social and cultural contexts. The curricula must be constructed in a manner that it is internationally accepted to match knowledge and skills round the globe. The knowledge, skills, and technology of any country determine its economic process indirectly (Patrinos, 2016). Rothstein (2010) says that reforms on education should be focused to bring change in economic and social policies to reduce the gaps that student face outside the classroom. Evaluation of curriculum should be done periodically and should be modified to reflect the changing times. Information and experiences should be added, deleted, or adjusted to reflect prevailing beliefs or findings. Every country has its own association that takes care of the standard education. In India, the National Board of Accreditation (NBA) suggests a procedure for quality assessment in the technical education sector. NBA was established during in, 1994. Following are the targets of NBA:

- To make the factors cohesive for assessment of quality of education
- To focus more on parameters to quantitatively assess criteria and assign appropriate program-specific weights
- To provide procedure authorization by well-designed test runs
- To determine benchmarks for important judgements

To test the standard of education imparted is one objective of NBA; others are the competence of the graduates and their relevance to the technical staffing needs in industry. Timely quality checks and improvement are required because the technical education sector in India is expanding quickly. The International Organization for Standardization (ISO) movement in education has highlighted the necessity for accrediting programs. They have technically qualified human resource and play an important role (National Board of Accreditation, India, 2009).

There are challenges to recommending subjects and learning areas in the curriculum. But it must develop student competency in areas such as the following (Stabback, 2016):

- best ways of communication
- collaboration for better understanding
- involvement in critical thinking

- ability in problem solving
- reflection of creativity in work
- managing diversity and versatility
- ability of learning

4.2.9 Use of Six Sigma for Quality Education

Six Sigma is a mathematical approximation of the quality level of products and services in terms of defective parts per million which is measured before products reach the customers. Six Sigma is used to achieve the goals of "higher, cheaper, and faster." This system, introduced by Motorola in the USA, was a large success in quality-improvement techniques. This procedure continues for an indefinite period and is re-evaluated and reformed until the product reaches the proper standard. This is also applicable for education sector to bring continuous improvement of industrial quality control to engineering education to feature more value to that (Wulf, 2002).

Six Sigma methodology is systematically improving process quality and productivity. It's going to be employed in case of education to lift its quality. Six sigma focuses on following areas: to improve people, to improve processes, to enhance products, and to improve materials. Six sigma helps to improve people by developing skills, capability, and problem-solving using "improved processes." It helps in removing inconsistency and improves stability in the system and matches method to customer needs (Devanathan, 2011).There are a few steps in Six Sigma project to be adapted, which are characterized as DMAIC, described below:

- **D**efine means defining the project, application area, improvement required, benefits to customers, etc.
- **M**easure means to spot key customer requirements, quality and process characteristics, etc.
- **A**nalyze means analysis of data and finding variables that have the best impact on the standard metrics of interest within the project
- **I**mprove means increasing high level of quality and implementing corrective and improved actions for best results.
- **C**ontrol means monitoring key process through control charts and quality attributes, documentation, etc.

4.2.10 Learning Based on Case Study: Addition in Curricula

Case studies are used as a teaching tool in which certain situations of organization are given to students. Students give suggestions through research and reflective discussion. In a case study, students are encouraged to think deeply for the solutions in an exceedingly creative manner by applying previously acquired skills. Case-study pedagogy is highly recommended for students because it involves analyzing the problem and developing analytical skill. The curricula should involve case studies as an activity in parallel to the concept teaching (Bonney, 2015). The case-study methodology must be included in curricula. It cultivates interest in teaching and involves students in discussion resolving tricky questions. Case studies develop the higher levels of Bloom's Taxonomy of cognitive learning by solving problems. The case-study methods help in analysis and evaluation of data (Anderson and Krathwohl, 2000). Case-study methodology enables interdisciplinary learning and relates with actual issues in organizations (Bonney, 2013). It motivates participation and performance of students. Case studies are used as a pedagogical framework of engineering ethics education in the USA and are supplemented by moral theory (Herkert, 2010) for the students. Ethical and interactive cases are widely disseminated in textbooks and online. Curriculum models in the USA have enriched courses in for overall development of young scholars. There is an integration of engineering ethics with science, technology, and society material. It's a challenge for engineering faculty to accept greater responsibility for engineering ethics education. They feel that case studies should be used irrespective of the teaching approach, in most engineering ethics curriculum (Herkert, 2010). The American Society for Engineering Education (ASEE) believes that engineering has a large and growing impact on society and must be equipped to fulfill their ethical duties toward society.

4.3 DISCUSSION

A systematic approach in engineering curricula emphasizes teaching ethical issues by incorporating a series of contextualized activities into the engineering curriculum. This approach may standardize the performance of engineers and needs to adhere to the very best principles of ethical conduct. Engineers need to perform professional duties considering the safety/protection, health, and welfare of the general public.

This process may strengthen the framework of engineering ethics curriculum and a process-driven research method. It will further help in identifying appropriate delivery strategies and instructional strategies. This ethical approach will enhance learning as well as developing engineering curriculum outcomes (Li and Fu, 2010). Before content is used in curriculum, its moral aspects must be judged by analyzing facts and its application. The theories of ethics developed by engineering societies may be used as a basis for rational selection for the courses (National Apprenticeship Promotion Scheme, 2019).

4.4 CONCLUSION

The change in curricula should be significant, bringing benefits in engineering education and technology competitiveness. The systematic educational approach will make engineers professionally competent. There is no doubt that technological progress and competitiveness of our nation will be enhanced by capability of the engineers. It should be based on ethical awareness and a mixture of research and be able to achieve excellence in professional education with innovations. The principles of ethics are well accepted by the society and follow the moral standards. An engineer who practices ethics is capable of helping society in a better way. As part of engineering research, ethical practices will help in robust decision making, policies and moral values. Needs- and motivation-driven technology is the primary propulsion for the nation's technological progress. The systematic educational curricula are going to be able to produce future creative leaders to satisfy the requirements of society and industry.

REFERENCES

Aizpun, M., D. Sandino, and I. Merideno. 2015. Developing students' aptitudes through university–industry collaboration. *Ingeniería e Investigación*, 35(3), 121–128. https://dx.doi.org/10.15446/ing.investig.v35n3.48188.

Anderson, L.W., and D. Krathwohl. 2000. A taxonomy for learning, teaching, and assessing: A revision of Bloom's Taxonomy of educational objectives, complete edition. White Plains, New York: Longman Publishing Group. http://dspace.vnbrims.org:13000/xmlui/handle/123456789/4570.

Anscombe, E. 1958. Modern moral philosophy, *Philosophy*, vol. 33, reprinted in Anscome, E. 1981. *Ethics, Religion and Politics*. Oxford: Blackwell.

Arvanitis, S., U. Kubli, and M. Woerter. 2008. University–industry knowledge and technology transfer in Switzerland: What university scientists think about co-operation with private enterprises. Research Policy, 37(10): 1865–1883. http://doi.org/10.1016/j.respol.2008.07.005.

Baruh, H. 2012. A need for change in engineering education. https://www.researchgate.net/publication/253731281_A_Need_for_Change_in_Engineering_Education.

Bonney, K.M. 2013. An argument and plan for promoting the teaching and learning of neglected tropical diseases. Journal of Microbiology & Biology Education, 14(2): 183–188. https://doi.org/10.1128/jmbe.v14i2.631.

Bonney, K.M. 2015. Case study teaching method improves student performance and perceptions of learning gains. Journal of Microbiology & Biology Education, 16(1): 21–28. https://doi.org/10.1128/jmbe.v16i1.846.

Bush, V. 1945. *Science: The Endless Frontier.* Washington, DC: Office of Scientific Research and Development (reprinted 1990 National Science Foundation).

Chhabra, T.N. 2016. *Industrial Management, Quality Control and Control Charts*, first ed., ISBN:978-93-85071-06-5, New Delhi: Sun India Publications.

Cruz, M.L., G.N. Saunders-Smits, and P. Groen. (2019): Evaluation of competency methods in engineering education: A systematic review, European Journal of Engineering Education, https://doi.org/10.1080/03043797.2019.1671810.

Damanpour, F., and S. Gopalakrishnan. 1998. Theories of organizational structure and innovation adoption: The role of environmental change, Journal of Engineering and Technology Management, 15(1): 1–24. https://doi.org/10.1016/S0923-4748(97)00029-5.

Dejmal, A. 2020. Key features of the learning organization, Chapter 9/Lesson 2, Strategic Human Resource Management. study.com.

Devanathan, S. 2011. A quality methodology for continuous improvement. http://aview.in/allevents/Six-Sigma-in-education-To-encourage-quality-education.php

Gibb, A. 2005. Towards the entrepreneurial university: Entrepreneurship education as a lever for change. National Council for Graduate Entrepreneurship Policy Paper #003. Birmingham: NCGE.

Hargadon, A., and R.I. Sutton. 2000. Building an innovation factory, Harvard Business Review, 78(3): 157–166.

Hegarty, C. 2006. It's not an exact science: Teaching entrepreneurship in Northern Ireland. Education + Training. 48(5), 322–335. https://doi.org/10.1108/00400910610677036.

Herkert, J.R. 2010. Engineering ethics education in the USA: Content, pedagogy and curriculum. European Journal of Engineering Education, 25(4): 303–313. https://doi.org/10.1080/03043790050200340

Hollister, S.C. 1949. Differentiating characteristics of an engineering curriculum, Engineering Education, ASEE, 291–294.

IES General Studies. 2017. Ethics in engineering profession. https://www.scribd.com/document/384129439/Ethics-in-Engineering-Profession-IES-General-Studies.

Kelly, S. 2011. Embedding enterprise education into the curriculum. University of Huddersfield. http://eprints.hud.ac.uk/id/eprint/9660/1/Embedding_Enterprise_Education_into_the_Curriculum.pdf

Kumar, M. 2015. https://issuu.com/udacademy/docs/ethics.pptx

Knowles, M. 1973. *The Adult Learner: A Neglected Species.* Houston: Gulf Publishing.

Lekhi, R. 2007. *Public Service Innovation: A Research Report for the Work Foundation's Knowledge Economy Programme.* London: The Work Foundation.

Li, J., and S. Fu. 2010. A systematic approach to engineering ethics education. Science and Engineering Ethics 18, 339–349. https://doi.org/10.1007/s11948-010-9249-8.

National Apprenticeship Promotion Scheme. 2019. Ministry of skill development & entrepreneurship Govt. of India. https://nsdcindia.org/sites/all/themes/ibees/pdf/naps-guidelines.pdf.

National Board of Accreditation, India. 2009. Evaluation guidelines for NBA accreditation of undergraduate engineering programmes. https://www.nbaind.org/Files/Guidelines%202009.pdf

National Council for Graduate Entrepreneurship. 2008. Developing entrepreneurial graduates: Putting entrepreneurship at the centre of higher education. Joint report of the National Council of Graduate Entrepreneurship, the Council for Industry and Higher Education, and the National Endowment for Science, Technology, and the Arts. Birmingham: NCGE.

National Society of Professional Engineers. 2019. nspe.org.

Organisation for Economic Co-operation and Development. 2016. *Education at a Glance 2016: OECD Indicators.* Paris: OECD Publishing.

Purzer, S., and J.P. Quintana-Cifuentes. 2019. Integrating engineering in K-12 science education: Spelling out the pedagogical, epistemological, and methodological arguments. Disciplinary and Interdisciplinary Science Education Research, 1, 13. https://doi.org/10.1186/s43031-019-0010-0.

Pahl, G., and W. Beitz. 1996. *Engineering Design: A Systematic Approach*, 2nd ed., trans. K. Wallace, L. Blessing, and F. Bauert. London: Springer-Verlag. https://doi.org/10.1557/S0883769400035776.

Patrinos, A.H. 2016. Why education matters for economic development. World Bank Blog. https://blogs.worldbank.org/education/why-education-matters-economic-development

Rajan, S.D. 2017. Incorporating ethics in engineering education. Journal of Engineering Education Transformations, 30(3): 164–171.

Rothstein, R. 2010. How to fix our schools: It's more complicated, and more work, than the Klein-Rhee "Manifesto" wants you to believe (Issue Brief #286). Washington, DC: Economic Policy Institute, ERIC Number: ED516800.

Santoro, M.D., and P. Saparito. 2003. The firm's trust in its university partner as a key mediator in advancing knowledge and new technologies. IEEE Transactions on Engineering Management, 50(3): 362–373. https://doi.org/10.1109/TEM.2003.817287.

Stabback, P. 2016. What Makes a Quality Curriculum? In-Progress Reflection No. 2 on "Current and Critical Issues in Curriculum and Learning." UNESCO International Bureau of Education [12102], [8],IBE/2016/WP/CD/02,2016.

Stitt, F.A., ed. 1999. *Ecological Design Handbook: Sustainable Strategies for Architecture, Landscape Architecture, Interior Design, and Planning.* New York: McGraw-Hill.

Standford, T.G., and D.A. Keating, 1998. An integrative approach to teaching and learning at the professional level for graduate engineers in industry. Paper presented at 1998 Annual Conference, Seattle, Washington. 10.18260/1-2--7224.

Tsai, W. 2001. Knowledge transfer in intraorganizational networks: Effects of network position and absorptive capacity on business unit innovation and performance. Academy of Management Journal, 44(5): 996–1004. http://doi.org/10.2307/3069443.

Wulf, W.A., and G.M. C. Fisher. 2002. A makeover for engineering education. Issues in Science and Technology 18(3): 35–39. https://www.jstor.org/stable/43314162.

WEBSITES

https://www.asme.org/topics-resources/content/approaches-to-teaching-engineering-ethics

https://www.nap.edu/read/11083/chapter/10 [accessed Jul 08 2020].

https://www.tutorialspoint.com/engineering_ethics/engineering_ethics_introduction.htm [accessed May 16 2020].

https://www.researchgate.net/publication/276102813_Promoting_Creativity_and_Innovation_in_Engineering_Education [accessed Jul 06 2020].

https://www.nspe.org/resources/ethics/code-ethics/ [accessed 02 June 2020]

https://iesgeneralstudies.com/ethics-in-engineering-profession/ [accessed 02 June 2020]

http://www.oecd.org/education/ceri/GEIS2016-Background-document.pdf/ [accessed 01 July 2020]

https://www.nbaind.org/Files/Guidelines%202009.pdf./an-integrative-approach-to-teaching-and-learning-at-the-professional-level-for-graduate-engineers-in-industry%20(1).pdf. [accessed 20 July 2020].

Section III

Tools and Methodology

5 Online Teaching in Universities in Developing Countries

Theory and Lessons

J. K. Tanui and F. M. Mwema
Dedan Kimathi University of Technology
Dedan Kimathi, Nyeri, Kenya

CONTENTS

DOI: 10.1201/9781003083160-8

5.1 INTRODUCTION

Higher learning institutions in developing countries still have many challenges with regard to offering quality education. The main challenge facing these institutions is the lack of human, physical and technical resources for delivering the services. However, in Kenya for instance, the government has made a great effort to improve the quality of learning in these institutions. The government is increasing the financial allocation for building infrastructure and facilities required by these institutions. The institutions have also partnered and collaborated with other universities and colleges in developed countries. Through such collaborations, the institutions in the developing world have been able to benefit in terms of research activities, staff capacity building through staff mobility program, and student exchange programs. The staff and students gain by using the laboratory equipment in their partnering institutions in the developed countries.

Most of the universities in developing countries offer classical engineering programs such as mechanical engineering, chemical engineering, electrical engineering and civil engineering. In addition to these main courses, sub disciplines and other engineering disciplines, such as mechatronics engineering, geospatial engineering, biomedical engineering, biomechanical engineering, telecommunication engineering, etc., are also offered. A learner in these courses requires a good theoretical background, foundation knowledge and practical skills. This is normally achieved when the number of students in a class is low, typically between 30 and 40. The lower the number the better the learning activity since the teacher is able to consider the learning need of each learner. However, this is sometime not the case in developing countries since the number of students in a class is higher, usually about 100 or above [1].

The main mode of teaching engineering courses in developing countries is the use of traditional teaching methods such as the lecture approach, tutorial-oriented, practical-oriented, project-based learning, etc. In the recent past, the use of technology such as instructional videos, computer tools, flipped classroom, virtual laboratory, among others, has gained popularity in many universities. However, implementation of these methods were still at development and piloting stage before the onset of Covid-19. The Covid-19 pandemic changed the perspective of learning in universities from face-to-face to online learning. Due to necessity and the need to stay financially afloat, even institutions that were not prepared have embraced this technology. Therefore, the use of technology is inevitable for the survival of institutions in future.

5.2 TEACHING OF ENGINEERING IN DEVELOPING COUNTRIES

It is the desire of each institution in the developing countries to produce and graduate learners who have the required skills to solve the current challenges faced by these countries. This is achieved by having well-developed curriculum and effective teaching. For many years, engineering courses, just like any other courses, have been taught at the university through lecture approach [2]. While this is the most adopted technique, the learner is able to acquire only three levels of Bloom's taxonomy: knowledge, comprehension and application [3]. The desire to improve the quality of graduates has led to the adoption of more effective teaching methods.

These methods include tutorial-oriented, practical-oriented, project-based learning, use of technology, etc.

The resources, common methods of learning and assessment in the universities in developing countries are discussed in subsequent subsections.

5.2.1 Resources

Successful implementation of an engineering course requires enough resources in terms of both human and physical facilities.

Universities in most developing countries still face a lot of challenges in human resources. According to a university statistics report by the Commission for University Education (CUE), in Kenya [1], Kenyan universities produce about 700 PhDs annually which is far fewer than the targeted number. Every year, the PhD graduands account for just 1% of the total graduands. Thus there is huge shortage of staff with PhDs in Kenyan universities. The few PhD holders are overburdened. They offer their services across several universities in terms of part-time appointments and adjunct professors. In addition, most of them take up the lucrative positions in the university management. The huge gap left is filled by staff who hold master's degrees. These staff teach as tutorial fellows under the supervision of staff with PhDs.

On the other hand, universities in most developing countries are not fully equipped with teaching facilities such as the laboratory equipment, library resources and teaching aids. The governments in these countries always try to avail the minimum teaching facilities required in order for the universities to achieve quality education. The few facilities available are of the current state of the art in their respective field of application since most of these universities are relatively new and these resources have just been acquired.

5.2.2 Common Learning Methods

5.2.2.1 Lecture Method

Lecture is the most common method of teaching courses in universities. Its efficiency and effectiveness are greatly dependent on the instructor, content and preparation of the instructor. When these factors are well set, then it is an efficient and effective way of teaching and learning. In most cases preparation of a lecture is left to the discretion of a professor. Usually, there is no external control, regulation or moderation of the lecture prepared by a professor. In a case where the lecture is not well prepared, the learners are greatly disadvantaged by this approach.

A good lecture is one in which not all content is covered by the professor. A great responsibility should be left to the student. For instance, the instructor should focus on the main principles, key points, general themes, difficult topics, etc. On the other hand, content that is easy to read and understand should be left as homework for the students. Also, when experiments are done, general procedures should be addressed by the lecture and fine details left to the students. A proper guide on how to select the content of the lecture is presented by Wankat and Oreovicz [2].

Delivery of the content through the lecture takes various forms, such as auditory, writing on the board, visual modes such diagrams, pictures, figures, drawings, etc. A lecture that is delivered entirely using one mode is not effective and may result in boredom. It is always advisable to have a mixture of delivery modes to have an interesting lecture.

The quality of a lecture is largely dependent on the pedagogical skills of the professor. In most universities of developing countries, professors are employed when they just obtained their PhD degrees without any pedagogical skills. They obtain these skills on job as they continue with their teaching. Most universities organize for short pedagogical modules that are normally held within the university. However, there are challenges in taking these modules because most professors believe that they already know how to teach and claim that they don't have time. Those professors who are willing to improve their pedagogical skills appreciate these modules. In fact, the feedback from the students for lectures by professors who have undertaken pedagogy modules shows that there is a great improvement in their presentation skills [4].

The main advantages and disadvantages of lecture are presented below.

Advantages of lecture

1. It is suitable for a big class or large number of students.
2. It does not require the use of technology.
3. The time required to prepare a lecture is less compared to other teaching methods.
4. It is appealing and suitable to students who are shy since it is difficult for the lecturer to ask all students questions.
5. It is flexible and versatile since other ways of teaching can easily be incorporated into a lecture.
6. The materials for lecture can be easily updated.
7. It is easy to get immediate feedback from students.
8. The professor has more control over the class as compared to other methods such as discussion.
9. It is an easy method of teaching in which new professors can easily adapt to it. It does not require special knowledge.

Disadvantages of lecture

1. It is easy for a lecturer to ignore the audience.
2. A professor may use inappropriate form of lecture.
3. Most lecturers do not change or update the learning materials use for lecture.
4. It is only limited to a few learning principles, i.e., knowledge, comprehensive and application. This problem is compounded when students are passive.
5. It does not have means of controlling preparedness. Thus there is lack of preparation for experienced professors and over preparedness for inexperienced professors.
6. The lecture can be boring if the professor has low voice, does not get feedback from learners and does not engage students.

7. It does not adapt for the learning pace of each student because it is controlled by the professor.
8. It may be extremely stressful for some professors.
9. The learning is inadequate, especially if the learning objectives and materials are not provided to students.

5.2.2.2 Tutorial-Oriented Approach

The tutorial-oriented approach is a form of learning that involves taking students through tutorial exercises in addition to lectures. This is student-centered learning in which the tutor ensures that each student's learning need is considered. It is a suitable form of learning for most engineering courses, which are analytical in nature and have a lot of calculation and computation. This form of learning is not applied independently without another method. It is meant to complement the other method of learning. When a lecture approach is combined with tutorial approach, then the former offers the fundamental principles of the subject while the latter assist the students to grasp the concept and the application part. The main goal of tutorial-based learning is to make the students be better problem solvers.

Enough teaching staff is required in order to ensure a successful implementation of tutorial-oriented learning. The class is broken into small groups which are handled differently. This means one person tutors them at different times or different persons tutoring at the same time. In most developing countries [5], tutorial classes are handled by post graduate students, who are normally referred to as graduate assistants (GAs), teaching assistants (TAs) or tutorial fellows (TFs).

The main advantage of tutorial-based learning is the benefit to students, who are able to have better understanding of the subject. The students are also able to appreciate the importance of the course through solving real and practical problems in the tutorial classes. In addition, weaker students are able to learn at their own pace since they receive individual attention. Students are also able to ask more questions that would not have been asked in the ordinary lecture class since it is a more interactive session. While this form of learning offers a great benefit to students, its main disadvantage is the cost involved. This is with respect to staff and teaching space requirement. Another disadvantage is that the students always want the solution to their problems, and they might not be very keen on the overall principle governing the analysis of the problem.

5.2.2.3 Practical-Oriented Approach

Practical-oriented approach is a method of learning which involves carrying out practical or laboratory exercise in order to understand a theory or a phenomenon. Most of the engineering courses requires this method of learning. There are courses that are implemented purely through practical sessions. However, most of the courses require some sessions of practice in addition to lecture and tutorials.

The main goal of the practical-oriented learning approach is to equip learners with technical skills that will help them in execution of their professional work [6]. Through several practical sessions, the students are able to gain lifelong skills that are not easily forgotten. Some laboratory exercises are meant to understand a fundamental theory or principle in the subject. In this kind of practical, the main objective

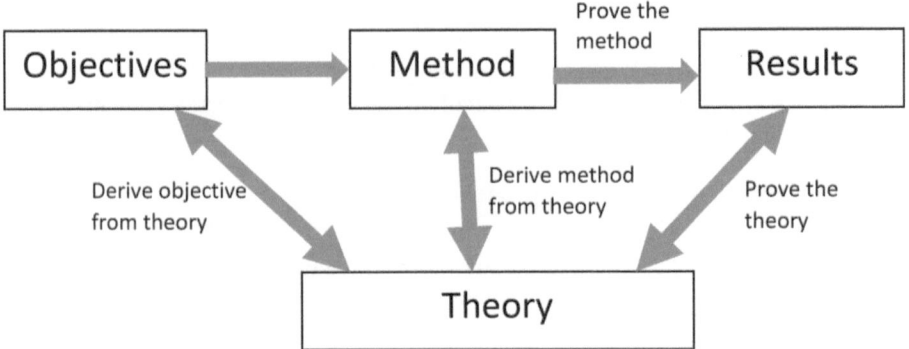

FIGURE 5.1 Interaction between the objectives, method and results of the practical and the theory.

is to prove the theory or method, as shown in Figure 5.1. The learners carry out analysis from the practical results data and compare them with theoretical analysis.

In order to achieve effective learning through practical-oriented approach, a small group of students, typically three to five students, is required for a given session. Figure 5.2 shows an example of a group of students in one of their practical session at mechanical engineering laboratory of Dedan Kimathi University of Technology. Because students are grouped into small numbers, multiple pieces of laboratory equipment and several technical staff are required. This increases the cost of training engineering students.

FIGURE 5.2 A group of students in a practical session at mechanical engineering laboratory of Dedan Kimathi University of Technology.

The main benefit of this method of learning is the ability to produce students that possess technical skills. On the other hand, the main disadvantage is the cost associated with equipment and staff. In addition, since it is implemented by grouping students, there are some learners who are not participating and leave the work to their colleagues.

5.2.2.4 Project-Based Approach

Project-based learning is a method of learning in which students are given projects that cover the whole spectrum of what they are supposed to learn. Courses that utilize this method of learning are mostly those that have a design aspect. The concept and theory are first introduced to the students before embarking on project implementation. Learning through the project-based approach is a high-level method that involves synthesis and combination of different concepts in order to come up with a system, component or product [7]. The concepts required may include, but are not limited to, scientific, mathematical and engineering principles [8, 9]. The theory is practically applied in a project in order to provide a solution.

Most of the projects are implemented in groups of two to five students. However, small projects are done individually. In the case of group projects, each student is assigned a specific task. The role of the lecturer is to give guidance on the execution of the project. This includes guidance on the goal, objectives, methods and outcome of the project. In addition, in most cases the lecturer has more experience and knows the time limit to achieve a particular objective of the project. Thus he/she advises the student accordingly.

Various types of projects can be undertaken by students, such as creative versus structured, minor versus major, disciplinary versus interdisciplinary, etc. In creative projects, the students come up with their own ideas. Even though they are open to a wide range of imagination, they are limited to a particular theme or topic. Structured projects are ones that follow a particular guidance or layout specified in the syllabus. The project may also be classified according to the size, i.e., a minor project that covers a small section of the course and a major project that covers a wide or the whole range of the course. The project may also apply to a particular discipline such as mechanical engineering, electrical engineering, etc., and is referred to as disciplinary project. On the other hand, a project that requires the input of experts from different disciplines is known as interdisciplinary [10]. Such projects are usually interesting and mostly represent solution to problems that are encountered in real life.

5.2.3 Course Assessment Approaches

Different methods of assessment may be adopted in engineering courses. The assessment method adopted depends on the course content and requirements. The assessment mode may combine some or all of the following means of examination: continuous assessment tests (CATs), final examination, laboratory practical examination and assignment. First, there are courses that are assess purely through continuous assessment, for instance, design, workshop practices and manufacturing courses. These are courses that are practical in nature, and the students are assessed based on continuous practical assignments. Second, there are courses that are assessed through a final exam, CATs and assignments. These are courses that do not require laboratory examination and may include mathematics and theoretical courses.

Finally, there are courses that are assessed through a final examination, CATs, laboratory practicals and assignments. These are engineering courses that have a laboratory practical aspect. In all the three cases, the final examination accounts for the largest percentage, usually up to 70%. The remaining percentage is shared by CATs, assignments and practicals.

5.3   ADOPTION OF TECHNOLOGY IN TEACHING ENGINEERING

Over the recent years, learning in higher institution has gradually improved through the use of technology in delivering the content. The technology adopted ranges from use of television, videos, projectors, computers and audio records. These delivery media are used together with the method of teaching in order to deliver content to students. It is important to distinguished between teaching method and delivery media. Teaching methods, which were discussed in the previous section, use delivery media to pass the information to the learners.

The delivery tools are applicable both to the live classroom, where the lecturer is physically present in class, and to remote/distance/online learning, where the lecturer and the learners are not in the same location. While it is possible to conduct a live classroom without use of technology, it is impossible to have a remote/distance/online learning without use of technology. With the onset of the Covid-19 pandemic, most universities, both in developed and developing countries, shifted their learning from live classroom to online learning. Therefore, the use of technology in learning is important now and in future. The characteristics of these technologies are discussed in the following subsections.

5.3.1   VIDEOS

Instructional video is the most common means of delivering content in a remote/distance learning course [11]. It is a suitable means of teaching post graduate students who are also working in industries. This means of content delivery does not interfere with their careers because the video can be watched at any time convenient to the learner.

Use of video is a very good means of showing visual representation of a system. For instance, in a course in mechanical engineering, the video can show the welding process from start to finish with emphasis on thermal effect on the material and quality of the weld. In a manufacturing course, the video may be used to show the raw material before assembly, the processes of assembling and the end product. In this way, the learner gets a better understanding of the subject than if it were just a theoretical class. In fact, it has been shown that learning through well-designed video is more effective than using a lecture method [12]. The instructor may be present in the instructional video. Alternatively, the video may not have the instructor or even the her/his picture. Even though the presence of the instructor does not add any value to the video, it has been argued that it improves the visual attention of the learner [13].

Videos may be used to supplement laboratory practicals. This is used where the equipment is delicate or not available in the university. Videos can also be used for practicals that are too dangerous and risky to be handled by students. For example, a video of the combustion reaction in a jet engine can be shown to learners who do not assess the engine. Videos can also be used to replace an industrial visit. In this case, a video of the factory and the processes carried out within it is taken. By watching the video, the learner gets the practical aspect of their studies without the cost and time associated with an industrial visit.

5.3.2 FLIPPED CLASSROOM

The flipped classroom concept is a mode of teaching whereby students learn the content at home using videos and the application part is done in the university [14, 15]. This concept has been developed to overcome the challenges associated with the classical lecture method, such as varied individual learning speed, individual learning style and lack of time for intensive advice to the students. Through the flipped classroom, there is more time available for learners to be trained by lecturers. In addition, students are able to learn the content individually at their own pace. Teaching using the flipped classroom improves the learning performance of the learners [16].

A successful implementation of flipped classroom requires the following key components:

a. **Instructional videos** – video of a lecture is recorded and sent to students either directly or through various online platforms. In order to be more effective, the videos should cover a small topic and be short, typically less than ten minutes.
b. **Face-to-face sessions** – during a face-to-face session, knowledge is deepened through discussion, answering and clarification of questions that arise, laboratory work, case analysis, peer teaching, brainstorming, etc. A face-to-face session involves three stages, namely, the input stage, the processing stage and the output stage. In the input stage, the lecturer briefly introduces the topic of discussion by explaining the background information. The processing stage is where the students broaden their knowledge through active participation, topic exploration, idea generation and brainstorming. This stage is carried out individually or in groups. The processing stage is led by students, and the lecturer act as the facilitator or moderator and provides support. The output stage is where the students present their work. The students are given feedback from their colleagues and the lecturer. This may be achieved in terms of posters, oral presentation, videos or written reports.
c. **Mini-tests** – these are test that are administered in each face-to-face session. They are meant to encourage and motivate students to learn the content in the video.

5.3.3 COMPUTERS

A computer is a resourceful tool that can be used to generate lecture notes and store lecture material and can be used directly during lecture delivery, with or without projectors. But more important is the specific instructional software that is installed on the computers.

The most common software used in engineering courses include AutoCAD, Inventor, SolidWorks, MATLAB, Ansys, etc. The use of computer packages, particularly spreadsheets, for calculation is mostly preferred by both the lecturers and the students, since they are user friendly [17]. Thus, it has provided a middle ground for hand calculators and computer programming.

The use of spreadsheets is applied in all engineering disciplines. The solution of problems may also be achieved through programming using equation-solving programs such as MATLAB, MathCAD, Fortran, Visual C++, etc. The use of these softwares is relatively hard compared to spreadsheets. It requires that the student knows how to write the equations and the programming language.

Teaching of design courses in mechanical engineering requires the use of CAD software such as Inventor, SolidWorks, AutoCAD, etc. They are very useful and have replaced manual drawing or drafting. Advantages of using CAD software include increased accuracy, possibility of 2D and 3D representation, connection to CAM software for production, less time, availability of standard templates, etc. However, it has a few limitations, such as computer breakdown, high cost of software installation, unsecure systems, among others.

5.3.4 VIRTUAL LABORATORY

Virtual laboratory is an online computational tool that simulates a given scenario and provides results. Virtual laboratory is applicable to all disciplines, such as art, literature, science, business and engineering. In engineering, virtual laboratories are program codes that enable students to interact with virtual machines, test different parameters and view results. The main goal of the virtual laboratory is to enable student to perform experiments when they are not physically present in location of the equipment [18, 19]. It is suitable for courses that have a practical component and are offered online. The Covid-19 pandemic changed the aspect of learning in most universities, which are now adopting online learning, and most are piloting the use of virtual laboratory [20].

The main benefit of a virtual laboratory is reduced cost of buying equipment. Even though the cost of purchasing virtual laboratory software is high, it is much cheaper than buying an actual machine. In addition, there is no physical space requirement. Experiments conducted on virtual laboratories are safe. There is no damage to equipment or injury to learners. It is easier to try out various scenarios and combinations of experimental parameters using a virtual laboratory than a real machine. It also helps to build the confidence in students since there is no handling of dangerous substances or working in restricted laboratory environment. The learners have a chance to repeat the experiments as many times as they want.

5.4 INTEGRATION OF BLOOM'S TAXONOMY INTO TECHNOLOGY IN TEACHING ENGINEERING

All levels of learning may be achieved when using technology to teach engineering courses by integrating Bloom's taxonomy. The flipped classroom concept is a perfect example of a technology that has all the levels of learning considered, as shown in Figure 5.3.

The learner understands the content in the video used in the flipped classroom concept by simply watching it. The act of watching the video enables the learner

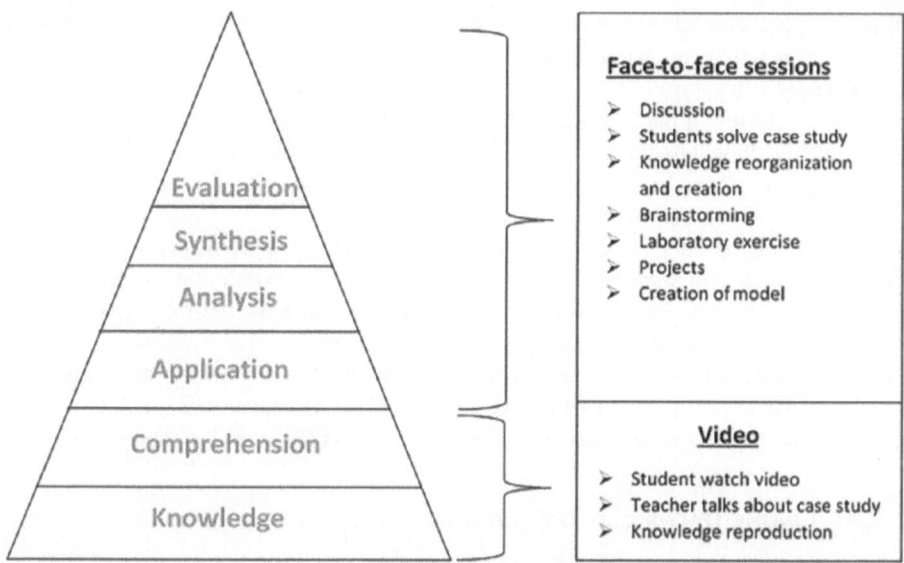

FIGURE 5.3   Bloom's taxonomy integrated into flipped classroom concept.

to reproduce the knowledge contained in it. The teacher might talk about a case study, explain a theory or illustrate a concept in the video. The students are able to understand what is presented in the video. Thus watching the video represents the lower-order thinking skills of knowledge and comprehension in Bloom's taxonomy.

Higher-order thinking skills of Bloom's taxonomy are achieved through class activities in the face-to-face session. In this session, the learners apply the formula given in the video to solve typical problems. They also use the knowledge given in the video to solve the case study on their own. By doing these activities, which are guided by a teacher, they are satisfying the application and analysis part of Bloom's taxonomy. Furthermore, the highest levels of Bloom's taxonomy, i.e., synthesis and evaluation, may be achieved if the students do a practical or theoretical project on the basis of knowledge of the video. Also they may give a theoretically based profound opinion for a given solution or model.

It is not possible to achieve all levels of learning without having a face-to-face session. For courses that are fully implemented online, it is also possible to have an equivalent of face-to-face session by use of platforms that allow interactive sessions. The most common available online learning and meeting platforms such as Zoom, BigBlueButton, Jamboard, MURAL, etc., have this provision.

5.5   A CASE DESCRIPTION OF ONLINE TEACHING

5.5.1   Description of the Case Study

In this section, the experiences of the authors involved in online teaching in a developing country, Kenya, are presented as a case study. Due to Covid-19, most universities in African and other developing countries in Asia have adopted an online mode of instruction to their students. DeKUT is one of the universities in Africa that has capitalized on online platforms to ensure teaching and learning during the Covid-19 lockdown. DeKUT is a public university duly registered and recognized by the Commission of University Education (CUE) of Kenya. The university was chartered in 2012 and focuses on training in the fields of science and technology. It offers engineering, technology in education, business, science, tourism, and health-based courses. In the school of engineering, where the authors are based, the university was able to enroll, orient and teach freshmen during the May–August semester in 2020 (Covid-19 pandemic period). During the same period, the school was able to teach and assess second-, third-, fourth-, and fifth-year students. The authors of the chapter are based in the department of mechanical engineering and were actively involved in teaching mechanical engineering courses. Generally, the courses in the department involve coursework, laboratory and project activities. The first author was involved in teaching coursework and laboratory-based courses in thermodynamics, while the second author was involved in teaching a purely project course (final year project 1) to fifth-year students.

5.5.2 Online Teaching by the Authors

During the Covid-19 lockdown, teaching at DeKUT was 100% online for the May–August semester. In this section, the implementation of two courses in mechanical engineering field/laboratory/coursework and project-based is described as a case study for online teaching in developing countries. The teaching was conducted through the BigBlueButton(BBB) and Moodle platforms. For each course, the following general procedure (Figure 5.4) for preparation and handling of the online classes was followed. As shown, the lectures involved three main tasks, namely scheduling, live lecture and after-lecture activities. The first task is to schedule the lecture on the online platform and provide students with details for accessing the lecture. The lecture should be created 24 hours prior to the live meeting and shared to all the students

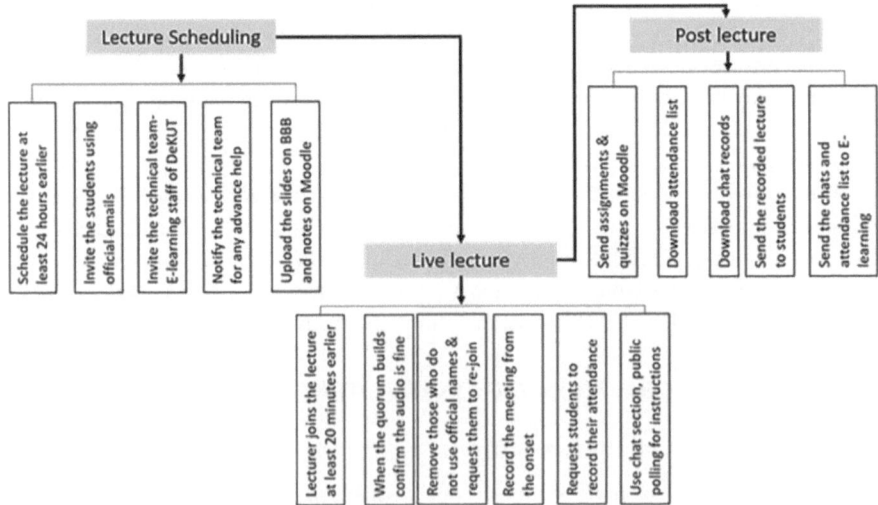

FIGURE 5.4   The three main tasks for online lectures prescribed for BigBlueButton.

in advance. During the scheduling the lecturer/instructor uploads all the necessary materials on the online platforms. The second task involves the live lecture, in which the virtual interaction between the instructor and students takes place. The most important aspects of this task are to ensure that students join the lecture with their official names and that the lecture is recorded from the beginning. Students should also write their names and identification numbers in the attendance list created by the lecturer on the platform. During the lecture, the students' microphones remain mute to avoid disruption of the class, and students can ask questions only via the chat function of the platform. Finally, after the meeting the lecturer should download all the chats and attendance list and share it with the quality assurance office. The lecturer should also download the lecture link and share it with all the students for future reference. The lecturer may also provide assignments or quizzes to the students via the Moodle platform to assess their understanding of the lecture concepts.

During teaching of the course-based unit, the most important aspect is the preparation of the presentation to the students. Usually lecturers use Microsoft® PowerPoint to prepare slide presentations for online lectures. To prepare a good slide presentation for online lectures, the following should be taken into account:

1. The presentation should utilize as many visuals as possible and reduce on the use of text as much as possible.
2. The visuals should be clear and easy to interpret to the audience. Avoid crowding too many figures and pictures on one slide.
3. Where necessary, sketch the figures during the lecture rather than projecting them. This helps the student grasp the concept better and follow the lecture.
4. Equations should not be copied as pictures in the slides. It is advisable for the instructor to write the equations during the live lecture. This makes the virtual class closer to face-to-face lecture. It also enhances interaction between the students and the lecturer.
5. The use of animation in the slides is also encouraged as it creates curiosity and motivation for the students to follow the lecture. It also makes the lecture interesting.

In addition to slides, it is important to prepare short notes on the subject to be discussed during the virtual live lecture. As per the experience of the authors, the best approach is to prepare the notes and provide them to the students before the virtual meeting (flipped classroom concept). In this way, the students are able to study and master the key agenda of the meeting in advance. As discussed in Figure 5.3, this approach ensures that the online meeting on the BigBlueButton focuses on the higher aspects of the Bloom's taxonomy. In DeKUT, the lecturer is supposed to upload all the notes and study guide materials on Moodle platform for access by the students before the virtual meeting. Based on the experience gained by the authors, the following important points should be considered when preparing these study materials.

1. The notes should be short, easy to understand and self-explanatory.
2. The use of sketches, pictures and illustrations in such materials is very important.

3. The notes should be divided into different concepts and subjects; i.e. a piece of study material should not be provided to cover the entire course content.
4. The lecturer is encouraged to include short quizzes and exercise activities in the study material to assist the students in better understanding the content.
5. Avoid copying and pasting content from books and other sources. Well-thought-out and developed notes are easy for the student to understand and follow.

During the live lecture, especially for course-based units in engineering, the lecturer should highlight the key aspects of the shared notes in their presentation and engage the students as much as possible through the necessary provisions of the online platform used. For instance, on the BB Button, the lecturer can ask questions on the public chat and see who responds fast and correctly. The lecturer can also make use of the polling function of the platform to confirm that everyone is attentive to the proceedings of the class. To make the lecture interesting, the instructor may decide to solve a typical case study question for the students during the live meeting. For instance, the platform used in DeKUT allows the lecturer to scribble during live lecture just like on a normal white board in face-to-face learning.

In conclusion, some of the tips for a good virtual meeting on an engineering subject include the following.

1. Make the lecture as short as possible. A meeting of not more than 20 minutes was seen to be the most effective by the authors of this article.
2. Illustrate the concept of the subject through solving a typical problem during the virtual meeting.
3. After every three to five minutes, pause and get feedback from the students on the public chat or polling function of the online platform.
4. Give short quizzes on the platform, and allow students to solve them in real time to ensure they are engaged and not bored throughout the lecture.

5.5.3 ONLINE ASSESSMENT

Assessment is one of the challenging aspects of online teaching and learning. Administering a closed-book and controlled examination and other assessments is nearly impossible. However, there are various lessons for alternative assessment during an online teaching process. The best approach to implement assessment for online teaching is the use of open-book tests, assignments and quizzes. One of the approaches the authors of this article commonly adopted during this period was to prepare three sets of open-book tests and assignments. Then the assessments would be shared with the students in a random way; on marking the assessments, this approach was seen to curb copying among the students. The best way to prepare open-book tests and assignments for engineering and science courses is to generate application-based tasks for the students. In this way, the responses of the students to specific tasks were found (by the authors) to be unique and interesting. These assignments and tests can also be complemented by random class quizzes in which, during the live online lecture, the students are asked to respond to questions and marks awarded. The quiz approach gives the lecturer assurance of the active participation of the students.

5.5.4 SUCCESSES AND CHALLENGES OF ONLINE TEACHING

The Covid-19 pandemic led to breakdown of face-to-face teaching and learning in the universities in Kenya. However, while all learning institutions in the country were locked down by the government, the institution of the authors (DeKUT) implemented online learning for all the students in various programs in the university. Thus, the students in this institution were able to progress in their studies, contrary to their peers in the other universities. In fact, the institution was the first in the country to admit, orient, and teach first-year students. One of the unique features of online teaching as observed by the authors of this article was that students were more confident to ask any question to their lecturers. In face-to-face lectures, some students tend to shy from asking questions or seeking clarifications; however, since the student's face is hidden behind the computer screen, it was not an issue for the shy students. As earlier mentioned, the lectures were usually recorded and later shared with the students. This way, those students who did not make it to the live lecture could also access the same content later. This means that online teaching enhanced class attendance and participation as compared to the face-to-face learning.

As this was the first time the authors were involved in a fully online teaching session, they experienced several challenges. The technological challenges of network connection and use of various devices by the authors (lecturers) were significant. Since most lecturers would offer these classes from their homes, students at times would complain of poor signal(s) from the lecturers. The same problem (poor network signal) would be experienced by the students. Teaching computation-based courses requires scribbling/sketching by the lecturers – it was quite challenging during the early stages of the online teaching sessions for the authors. However, there are various tools that can be used to write just like on a normal whiteboard. Microsoft® OneNote is one of the tools the authors found very versatile for scribbling/sketching during live lectures. However, such tools require computers or tablets with touch screen and a writing pen. As earlier mentioned, the other challenge of online teaching is the lack of practical sessions, and this calls for lecturers' creativity and extra effort to provide the students with as many illustrations, case studies or video examples as possible to enhance their understanding of different concepts.

5.6 CONCLUSIONS

With the ever-changing world and occurrence of uncertainties in the modern times, institutions of learning need to embrace technology to survive and accomplish their mandate. The outbreak of Coronavirus (Covid-19) in 2019–2020 led to the shutdown of teaching and learning activities in most universities in developing countries. One of the obvious ways to offer teaching in universities during such pandemics is to use online teaching. In this article, the authors have presented some lessons inferred from their involvement in online teaching at DeKUT. Additionally, the chapter provides an overview of technology in engineering teaching with emphasis on a technology-enhanced flipped classroom. The use of technology in the flipped classroom is explained as an effective method of implementing online teaching. The successes and challenges of online teaching are also presented in the chapter. This article will be useful to other teachers/institutions to adopt/enhance/implement online teaching for continuity of their academic activities, even during pandemic times.

REFERENCES

1. Commission for University Education, "University statistics (2017/2018)," http://cue. or.ke/index.php/downloads/category/18-universities-data-0-3?download=205:2017-2018-university-statistics-report-approved-doc, 2019.
2. P. C. Wankat and F. S. Oreovicz, *Teaching Engineering*. McGraw-Hill, 1993.
3. M. Farashahi and M. Tajeddin, "Effectiveness of teaching methods in business education: A comparison study on the learning outcomes of lectures, case studies and simulations," *Int. J. Manag. Educ.*, vol. 16, no. 1, pp. 131–142, 2018.
4. A. M. Walder, "Pedagogical innovation in Canadian higher education: Professors' perspectives on its effects on teaching and learning," *Stud. Educ. Eval.*, vol. 54, pp. 71–82, 2017.
5. F. M. Nafukho, C. S. Wekullo, and M. H. Muyia, "Examining research productivity of faculty in selected leading public universities in Kenya," *Int. J. Educ. Dev.*, vol. 66, no. August 2018, pp. 44–51, 2019.
6. A. Pich-otero *et al.*, "Laboratory practical work as a technological process Augusto," *Biochem. Educ.*, vol. 26, pp. 281–285, 1998.
7. R. Kuppuswamy and D. Mhakure, "Project-based learning in an engineering-design course – developing mechanical- engineering graduates for the world of work," *Procedia CIRP*, vol. 91, pp. 565–570, 2020.
8. A. Sharma, H. Dutt, C. Naveen Venkat Sai, and S. M. Naik, "Impact of project based learning methodology in engineering," *Procedia Comput. Sci.*, vol. 172, pp. 922–926, 2020.
9. P. Guo, N. Saab, L. S. Post, and W. Admiraal, "A review of project-based learning in higher education: Student outcomes and measures," *Int. J. Educ. Res.*, vol. 102, no. April, p. 101586, 2020.
10. M. Ricaurte and A. Viloria, "Project-based learning as a strategy for multi-level training applied to undergraduate engineering students," *Educ. Chem. Eng.*, 2020.
11. M. Bétrancourt and K. Benetos, "Why and when does instructional video facilitate learning? A commentary to the special issue 'developments and trends in learning with instructional video,'" *Comput. Human Behav.*, vol. 89, pp. 471–475, 2018.
12. B. B. de Koning, V. Hoogerheide, and J. M. Boucheix, "Developments and trends in learning with instructional video," *Comput. Human Behav.*, vol. 89, pp. 395–398, 2018.
13. J. Wang and P. D. Antonenko, "Instructor presence in instructional video: Effects on visual attention, recall, and perceived learning," *Comput. Human Behav.*, vol. 71, pp. 79–89, 2017.
14. S. Guraya, "Combating the COVID-19 outbreak with a technology-driven e-flipped classroom model of educational transformation," *J. Taibah Univ. Med. Sci.*, vol. 15, no. 4, pp. 253–254, 2020.
15. P. Strelan, A. Osborn, and E. Palmer, "The flipped classroom: A meta-analysis of effects on student performance across disciplines and education levels," *Educ. Res. Rev.*, vol. 30, no. January, p. 100314, 2020.
16. G. Akçayır and M. Akçayır, "The flipped classroom: A review of its advantages and challenges," *Comput. Educ.*, vol. 126, no. July, pp. 334–345, 2018.
17. B. R. Lawson, K. R. Baker, S. G. Powell, and L. Foster-Johnson, "A comparison of spreadsheet users with different levels of experience," *Omega*, vol. 37, no. 3, pp. 579–590, 2009.
18. J. Grodotzki, T. R. Ortelt, and A. E. Tekkaya, "Remote and Virtual Labs for Engineering Education 4.0: Achievements of the ELLI project at the TU Dortmund University," *Procedia Manuf.*, vol. 26, pp. 1349–1360, 2018.
19. D. Liu, P. Valdiviezo-Díaz, G. Riofrio, Y. M. Sun, and R. Barba, "Integration of Virtual Labs into Science E-learning," *Procedia Comput. Sci.*, vol. 75, no. Vare, pp. 95–102, 2015.
20. J. Glassey and F. D. Magalhães, "Virtual labs – love them or hate them, they are likely to be used more in the future," *Educ. Chem. Eng.*, vol. 33, pp. 76–77, 2020.

6 Significance of Outcome-Based Assessment and Evaluation-Centered Performance Indicators for Continuous Quality Improvement of Engineering Education

*Didem Okutman Tas, Ebru Dulekgurgen,
Gulsum E. Zengin, Burcak Kaynak, and
Cigdem Yangin-Gomec*
Istanbul Technical University
Istanbul, Turkey

CONTENTS

DOI: 10.1201/9781003083160-9

6.1 INTRODUCTION: BACKGROUND AND DRIVING FORCES

The paradigm shift from teacher-centered to student-centered higher education and the expectations of the public and private sectors from higher education graduates have changed due to the needs of the evolving societies of the current era of "change," "quality," and "sustainability;" as well as the recent trends in higher education, such as internationalization, global challenge, and university rankings. These have all increased the popularity of accreditation, which favorably affects academic quality. Accreditation also contributes to the improvement of processes and practices in a program or institution (Dulekgurgen et al. 2018a, 2018b, 2018c, 2019a, 2019b, 2022; Ulker and Bakioglu 2019). Sin et al. (2017) reported accreditation as quality evidence for an institution or an undergraduate program. ABET – one of the non-profit, non-governmental, internationally acknowledged and prestigious quality assurance and accreditation (Q&A) agencies providing periodical third-party review and accreditation of undergraduate and graduate programs worldwide in the disciplines of "applied and natural sciences, computing, engineering, and engineering technology" – has proven its competence and strength as one of the first accrediting organizations to apply a philosophy of continuous improvement for accreditation and the first to present that process to scholarly evaluation (Akduman et al. 2001; Volkwein et al. 2007). Seeking and keeping acknowledgement of quality assurance of higher education, in other words, seeking and keeping accreditation of the programs by internationally recognized prestigious third-party agencies (e.g. the ABET) is a long-term, voluntary, and effort-intense process and is considered an important institutional asset, because periodical review leading to accreditation by an internationally recognized third-party QA&A agency holistically contributes sustainability of continuous improvement of education and learning processes, structures, and experiences for the primary constituents of higher education – the students and the instructors.

Program accreditation is one possible outcome of quality assurance process(es); however, its meaning and implications vary from country to country and subject to subject (Frank et al. 2012). Academic leaders at universities around the world prefer to maintain accreditation as an outcome of promoting globalization and internationalization of higher education (Barrett et al. 2020). Moreover, an accreditation process leading to better content of the programs may also assist in attracting better students and faculty members (Shafi et al. 2019). Ryan (2015) reported the necessity of having an international quality guarantee framework with acknowledgement and reciprocity. In this respect, assuring and sustaining the quality of education has vital importance; this could be achieved through a sequence of step-wise processes and procedures, primarily starting with self-evaluation of the program and then getting reviewed and evaluated by independent local and/or national QA&A agencies and/or by an internationally recognized accreditation organization; such as ABET. Such a multi-stage review and accreditation process generally starts with internal evaluation of the quality system before external evaluation by peer reviewers (Asif et al. 2013; Stura et al. 2019). For instance, in order to acquire ABET accreditation, a program needs to submit its self-study report (SSR), which is a detailed compilation of quantitative and qualitative assessment and evaluation of the program, highlighting its strengths and limitations. SSR provides a comprehensive overview of a program and all the information necessary for ABET's voluntary program evaluators (PEV) to review the program with regard to several preset criteria. External evaluation may result in full or partial compliance with the accreditation criteria, and in case of the latter, the programs are required to develop and implement remedial action plans, assess and evaluate the outputs of the remediation processes, and make use of those in a knowledge-based decision-making process for further fine-tuning of the programs, which then contribute to the continuous improvement of the programs (Dulekgurgen et al. 2018a, 2018b, 2019a, 2022). In this respect, a program needs to maintain/sustain the continuous improvement process through assessing and evaluating the level of attainment of the SOs, generating recommendations for change when needed, taking remedial actions/decisions, and implementing recommendations obtained from the results of the SO A&E process. ABET EAC's descriptions for the three critical concepts of the continuous improvement process are as follows (ABET EAC 2018; ABET EAC 2022):

> ***Student Outcomes*** *– Student outcomes describe what students are expected to know and be able to do by the time of graduation. These relate to the knowledge, skills, and behaviors that students acquire as they progress through the program.*
> ***Assessment*** *– Assessment is one or more processes that identify, collect, and prepare data to evaluate the attainment of student outcomes. Effective assessment uses relevant direct, indirect, quantitative and qualitative measures as appropriate to the outcome being measured. Appropriate sampling methods may be used as part of an assessment process.*
> ***Evaluation*** *– Evaluation is one or more processes for interpreting the data and evidence accumulated through assessment processes. Evaluation determines the extent to which student outcomes are being attained. Evaluation results in decisions and actions regarding program improvement.*

Assessment is considered a mutual interaction in which an instructor, a student, and peers can discuss the standards, criteria, and assessment experiences (Leenknecht et al. 2021). Three stages of evaluation have been reported: submission of self-evaluation documents, visits/reviews by external evaluators, and a review report preparation. In this context, student outcome assessment and evaluation (SO A&E) results need to be congruent in order to close the loop that is the complete cycle reflecting the efforts for continuous improvement, eventually to result in a successful accreditation (ABET 2022; ElAli 2017; Frazer 1992; Jacob and Boyter 2020).

In addition to the targeted genuine merits in regard to sustaining educational quality, accreditation might serve as a valuable asset for pursuing international academic recognition (Barrett et al. 2020; Frazer 1992; Morell 2010; Prados et al. 2005). Achieving quality standards in education and the process of continuous improvement acknowledged by accreditation is the main driving force for the higher education sector (Dayananda et al. 2020). Together with increased emphasis on learning outcomes and student practices, the foci shifted from curricular specifications to achieving the SOs and accountability, and significantly higher levels of achievement have been reported (Volkwein et al. 2007). Since accreditation plays a vital role, institutions have started to put more emphasis on SOs by means of accreditation

(Ulker and Bakioglu 2019). As stated above, ABET describes student outcomes as "what students are expected to know and be able to do by the time of graduation that are related with the knowledge, skills, and behaviors that students gain as they progress through the program." However, a deep and common understanding of the content and level of the underlying competences of the SOs is a prerequisite, and a program needs to determine and list the SOs that support its program educational objectives (PEOs). Moreover, any program seeking initial or re-accreditation is required to demonstrate that it is assessing and evaluating the extent to which the SOs are attained on a regular basis. Successful and satisfactory achievement of the SOs implies that the graduates are well equipped and prepared for entering their professional practice of engineering (ABET 2022; Strijbos et al. 2015).

In this respect, most of the international and national accreditation agencies have taken student outcome assessment a step further by identifying specific learning outcomes to be achieved by programs seeking accreditation and have described the undergraduate learning outcomes common for all baccalaureate degrees in engineering, regardless of engineering specialty (Volkwein et al. 2007). On the other hand, accreditation itself is a continuously evolving process by nature, incorporating lessons learned, best experiences, and improvements (Prados el at. 2005). Accordingly and as the most recent example of continuous improvement in the engineering accreditation criteria, the 11 SOs of the EAC of ABET (a-to-k), which were first described in the EC2000 (*aka ABET Engineering Criteria 2000*, 1995) and utilized for the past two decades, have recently evolved into a revised set of the following 7 SOs, which have been adopted as of the 2019–20 accreditation review cycle (ABET EAC 2018, ABET EAC 2022):

SO1 – an ability to identify, formulate, and solve complex engineering problems by applying principles of engineering, science, and mathematics

SO2 – an ability to apply engineering design to produce solutions that meet specified needs with consideration of public health, safety, and welfare, as well as global, cultural, social, environmental, and economic factors

SO3 – an ability to communicate effectively with a range of audiences

SO4 – an ability to recognize ethical and professional responsibilities in engineering situations and make informed judgments, which must consider the impact of engineering solutions in global, economic, environmental, and societal contexts

SO5 – an ability to function effectively on a team whose members together provide leadership, create a collaborative and inclusive environment, establish goals, plan tasks, and meet objectives

SO6 – an ability to develop and conduct appropriate experimentation, analyze and interpret data, and use engineering judgment to draw conclusions

SO7 – an ability to acquire and apply new knowledge as needed, using appropriate learning strategies

The above listed SOs are of particular importance due to the fact that two pillars of quality assurance and accreditation might be defined as keeping track of the level of attainment of the "student/learning outcomes" as the benchmark and "continuous quality improvement" as the core process (Dulekgurgen et al. 2018b, 2019a).

Moreover, it is also required to describe how the results of those processes are utilized in supporting continuous improvement of the programs. The true benefits of accreditation lie in the ongoing process of assessment and evaluation that lays the foundation for a program's success.

In this context and together with all eight engineering accreditation criteria (i.e., students, program educational objectives, student outcomes, continuous improvement, curriculum, faculty, facilities, institutional support) specified by the ABET EAC, the program assessment planning and continuous improvement strategy of any program seeking initial or re-accreditation should primarily focus on the following three core criteria (ABET 2022):

- *ABET EAC Criterion 2: Program Educational Objectives (PEOs)*
- *ABET EAC Criterion 3: Student Outcomes (SOs)*
- *ABET EAC Criterion 4: Continuous Improvement (CI)*

Because a SO A&E process is required to be demonstrated in the SSR of a program seeking accreditation/re-accreditation (i.e., especially the process enabling and the solid proofs demonstrating compliance with Criterion 4: Continuous Improvement), each program should be running a structured and a documented system for continuous improvement that actively and formally engages all of its constituents in the development, assessment, and improvement of academic proposals for providing accountability (Volkwein et al. 2007). Hence, a continuous improvement process (CIP) should be implemented by the programs in compliance with their institutional quality management and assurance strategies, which are also in line with the ABET's accreditation criteria. The programs must regularly use appropriate, documented processes for assessing and evaluating the extent to which the student outcomes have been attained due to the fact that when assessment does not match the targeted abilities, this can put students' learning at risk (Segers et al. 2009). Moreover, the results of these evaluations should be systematically used as input for the continuous improvement of the program. Moreover, a continuing effort is required to make assessments more standardized across the program, so that trends could be identified and evaluated in a systematic manner. An important consideration while planning the assessment and evaluation processes at SO-, individual course, four-year curriculum, and overall program levels to include as many faculty members as possible in those processes and also determine the degree to which they are involved (Dulekgurgen et al. 2019a, 2019b; Shafi et al. 2019). It is critically important and among the best practices to get involvement and contributions of the faculty members, yet it is also as critical to have good time- and effort-planning in this respect as the A&E processes are likely to increase workloads of the faculty members, especially for the first stages (Shafi et al. 2019).

Although structuring comprehensive A&E strategies for determining the level of achievement of SOs arise compelling, designing an effective curriculum is one of the challenging tasks for all engineering undergraduate programs. Engineering education, especially, plays a central role in the growing technology-based societies and in providing design principles. Teaching how to identify, formulate, and solve the problems in engineering design should be obviously recognized and should be well performed in curricula. Assessment and evaluation processes have a central role in curriculum development and a key role in the curriculum–learning/teaching–assessment triangle (Dym et al. 2005; Morell 2010; Orsmond et al. 2006; Passow and Passow 2017; Struyven et al. 2014).

In this context, students' perception of assessment is of particular importance, and engineering programs have started to compile their curricula with the outcomes stipulated by their respective criteria. Moreover, especially a self-assessment process performed by the students contributes to the formation of critical thinking of students, activates their self-awareness, increases their interest and motivation to learning, and encourages them to be independent subjects of the educational process (Dulekgurgen et al. 2022; Saribeyli 2018). Evaluation by students (i.e. evaluating the instructor, the courses' content, the classrooms or anything and anybody pertinent to the program) is a complex process relying on many factors and includes a set of methods and tools (Nicolaou and Atkinson 2019).

Because higher education curricula focus mainly on delivery/achievement of information, knowledge, and competencies, and on technical and soft skills required by each profession and their employers (Mintz and Tal 2014), curriculum revisions and improvements should be initiated by the feedback obtained from the external constituencies – primarily the employers and, in the case of engineering education, the industry and the related public and private sectors – together with the internal constituents. This means that the A&E process must be structured in a way to generate outputs and informed decisions for curriculum reviews and improvements targeting continuous improvement recommendations. In this context, one of the primary foci of development of a new/revised curriculum is satisfactory achievement of the SOs upon graduation and PEOs within three or five years after graduation, as also recommended and required by ABET, and engineering programs must define their PEOs to meet their constituents' needs. In addition, programs must determine certain targets for student learning achievement and measure their performances to show how well these objectives are being met (Volkwein et al. 2007). Among the constituents, faculty members play a vital role in accreditation/re-accreditation process of their institutions (Dulekgurgen et al. 2019a, 2019b). In this respect, a curriculum map of the programs needs to be constructed with the joint involvement of the entire faculty as a fundamental roadmap of the SO A&E process at program level (Aoudia and Abu-Alqahsi 2015; Dulekgurgen et al. 2022; Fowler et al. 2010; Yangin-Gomec et al. 2016, 2018). Faculty members need to discuss the pedagogical aspects that are involved in the reshaping of the existing curriculum, which could be applicable by the outcome-based education (OBE) approach. It is reported that OBE could have a significant positive effect on students and prepare them both for exciting careers and for successful self-expression in engineering. On the other hand, Jongbloed et al. (2018) reported that reliable information and transparency on the benefits that higher education institutions offer to their students, funders, and communities are key to their legitimacy, funding, and competitiveness. In this context, with the help of outcome-based assessment, transparency within the curriculum could be also increased. Since all the programs have the main targets to be achieved through the four-year curriculum; the degree to which programs succeed in enabling students to learn is assumed to present a more transparent, more appropriate, and more locally-differentiated yet globally-complying picture of quality (Das et al. 2017; Jongbloed et al. 2018; Orsmond et al. 2006).

Using multiple assessment tools and methods to achieve and verify an outcome is an important feature of effective assessment because the more tools used to assess a specific SO or course learning objective, the greater the likelihood that the assessment will be both valid and dependable. In this respect, effective assessment uses relevant direct, indirect, quantitative, and qualitative measures as appropriate to the outcome being measured (Dulekgurgen et al. 2018a, 2022; Felder and Brent 2003; Yangin-Gomec et al. 2016).

There is considerable and growing interest in quality assurance and accreditation (QA&A) in higher education – and engineering education in particular – on a national basis in Turkey, primarily starting by late '90s upon initiation of internationally recognized QA&A processes by some top-ranked public and private universities. These have been preferred by the top-ranked and highly qualified students in the country, offering numerous engineering undergraduate programs accredited by the EAC of ABET and paving the way to improve engineering education and contribute to the ever-demanding and dynamically changing needs of the engineering profession and to the future of sustainable and resilient global societies, through their graduates (Dulekgurgen et al. 2018b). In parallel to those initial and ongoing international QA&A efforts and the individual successes of some top-ranked public and private universities, QA&A efforts have also been intensified on a national basis during the past two decades. The Council of Higher Education (YÖK) of Turkey – an umbrella institute governing all universities in the country – has been encouraging QA&A efforts on a national basis as well to promote and support the improvement of the quality of higher education throughout the country (Öz 2005). Accordingly, the first national agency on QA&A in higher education of the country, MUDEK (Association for Evaluation and Accreditation of Engineering Programs) was structured late in the first decade of the 2000s. It became a full signatory member of the Washington Accord in the early 2010s, and continues its activities on evaluating and accrediting engineering undergraduate programs on demand on a national basis (MUDEK 2022). More recently, YOKAK (The Higher Education Quality Council of Turkey – THEQC) was founded in 2015 as the second national agency on QA&A in higher education in the country, reviewing and accrediting higher education institutes of the country on demand (YOKAK 2022).

While some private universities have benefited from offering programs accredited by international accreditation agencies for marketing and improving their national distinction and global competitiveness (Barrett et al. 2020), it is considered that establishment of the standards (i.e., under the themes of SOs and CIP, curriculum, faculty, facilities, equipment and supplies, and fiscal and administrative capacity), attracts attention of international students, especially stemming from lack of a national quality assurance systems in their own countries (Ramírez and Luu 2018, Wilkerson 2017). In this context, students have been frequently asking the following question during their higher education: "What was I supposed to gain from this course?" after completing their course. Hence, SOs – the "what" the students are supposed to acquire from any course unit – are considered to be the starting point of the process of planning the content delivery and assessments tools and methods. It is considered that assessment may motivate students to learn, direct their efforts toward effective learning, and enable them to learn from the assessment process and feedback when appropriately designed and implemented (Vu and Dall'Alba 2007).

In this context and as described and emphasized by many others (Biggs and Tang 2011; Gerritsen-van Leeuwenkamp et al. 2019; Kumpas-Lenk et al. 2018; van der Vleuten et al. 2017), Istanbul Technical University (ITU)'s Environmental Engineering Undergraduate Program (EEUP) also considers that the success and effectiveness of the environmental engineering education it offers to its undergraduate students depends not only on the delivered content and assessments but even more on the quality and sustainability of the assessment and evaluation processes. Therefore, ITU EEUP, being one of the 25 engineering undergraduate programs offered by ITU and accredited by ABET EAC (ABET 2022; ITU EED 2022), has been putting the major emphasis on assessment and evaluation of the student outcomes and the students' learning experiences (Dulekgurgen et al. 2016, 2018a, 2022; Yangin-Gomec 2016, 2018).

Hence, the aim of this study is to provide some insight and attempt to meaningfully contribute to the field of QA&A in higher education through presenting the most critical part of the long-range and successful QA&A efforts by the ITU EEUP; this is through presenting an edited compilation of the sections from the SSR of the program from the previous re-accreditation review cycle, covering the assessment, evaluation, and continuous improvement processes and results.

Specific exemplifications are also provided:

i. direct utilization of the performance indicators (PIs) in SO A&E
ii. the PI-breakdown–based assessment methodology and results
iii. the approach, methodology, tools and measures, implementations, assessments, remediations, re-assessment, short- and long-term results within the framework of assessment and evaluation at individual SO level (ABET EAC's legacy 11 SOs, also known as SOs a-to-k) and at course level, as well as at program level through the senior-year compulsory cap-stone engineering design course.

6.2 METHODOLOGY

The first half of this current study includes an edited compilation of the foci-sections from the ITU EEUP's SSR prepared for a previous comprehensive re-accreditation review cycle, encompassing the assessment, evaluation, and continuous improvement processes and results. Those outline the structured, iterative, and periodic processes for assessment and evaluation at individual SOs and course levels, as well as at the overall program level; they also demonstrate students' perceptions and instructors' implementations and A&E of the program. The overall goal for such an A&E scheme and process flow is to enhance the quality of the program, the related processes, and eventually the environmental engineering education offered by the ITU EEUP, and thus to prepare its students for entering the workforce as fresh graduates fully equipped with the required technical and soft skills, knowledge, and competencies (Dulekgurgen et al. 2018b, 2018c). The SO assessment approach relying on direct and indirect evidences of students' learning as they progress in the program includes the following tools and methods as indicated in Figure 6.1.

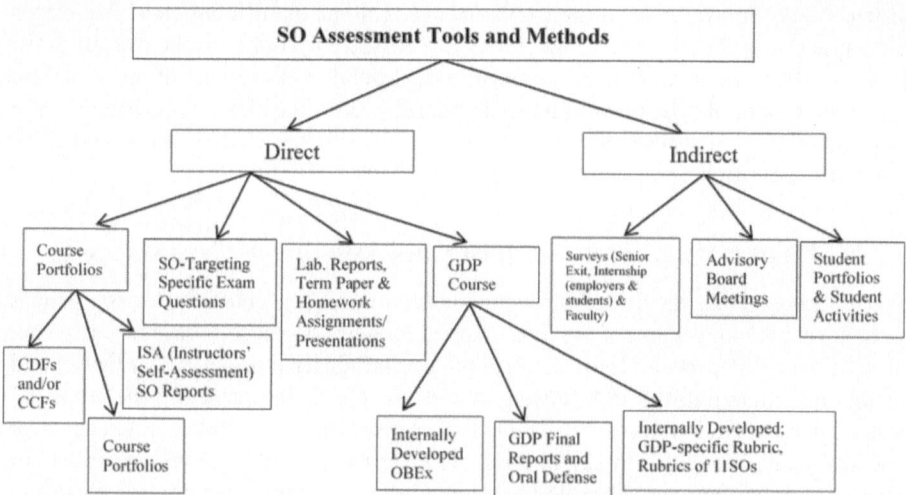

FIGURE 6.1   Direct and indirect assessment tools utilized by ITU EEUP for assessing the level of achievement of the SOs (ABET EAC's legacy 11 SOs: also known as SO a-to-k).

Detailed descriptions of the direct assessment tools and methods used in this A&E frame are given below, and additional information is available at the "Accreditation Info: ABET Accreditation" link of the publically accessible official web page of the department: https://cevre.itu.edu.tr/en/education/accreditation/abet-accreditation (ITU EED 2022):

6.2.1   COURSE PORTFOLIOS

The course portfolios (CPs) of compulsory courses were collected at the end of each semester by the assigned committee of the program and included the following contents for assessment of the student outcomes:

- course description form (CDF) and/or course catalogue form (CCF)
- course portfolio (CP)
- instructors' self-assessment reports (ISA): SO assessment reports prepared by the instructors of the compulsory courses mapped to each one of the 11 SOs, and using, for example, the related rubrics (including the PIs) prepared by 11 faculty members and supporting teaching assistants of the ITU EEUP; the former were officially assigned as the SO coordinators in charge of collecting, archiving, assessing, and evaluating all ISA reports submitted by the end of each term, and preparing an "overall SO A&E term report" to be submitted to the relevant governing/managing/coordinating parties within the department (i.e., department's administration, etc.).

6.2.2   SO-TARGETING SPECIFIC EXAM QUESTIONS

Instructors used specific questions in the quizzes, midterms, and/or final exams for assessment of level of achievement of the SOs mapped to their compulsory courses, which should be available in their CPs. Hence, neither the official end-of-semester letter grades (or the corresponding grades in 100-scale system) nor the overall exam grades of the students were used to measure the level of achievement of the SOs. The scores they obtained from SO-targeted specific exam questions/assignments were used– in other words those of the students' aggregating knowledge, skills, and competencies (Gerritsen-van Leeuwenkamp et al. 2019).

6.2.3 Laboratory Reports and Term Paper Assignments/Presentations

Instructors used lab reports and term paper assignments/presentations for assessment of the level of achievement of the SOs mapped to their compulsory courses including lab sessions and/or those requiring term papers and/or term projects. All the related documents, including the ISA reports as well as those demonstrating the students' performances (graded versions of selected originals or hard and/or soft copies of students' exams sheets, quizzes, reports, etc.), were provided by faculty members in the course portfolios and submitted to the relevant governing/coordinating parties at the end of the semester.

6.2.4 Homework Assignments

Instructors used homework assignments for assessment of the level of achievement of the SOs mapped to their compulsory courses requiring homework assignments. All the proofs of those were managed the same way as for the lab reports and term papers/projects, explained above.

6.2.5 Tools for the Graduation Design Project (GDP) Course

Compulsory engineering design courses have a particular importance in the curriculum due to their vital roles in improving students' aggregating and advancing knowledge, skills, abilities, and competencies related to problem/project-based learning (PBL) and in conveying program/content-specific knowledge to the students, as well as developing and sharpening their engineering design skills. The "CEV492/E Graduation Design Project" GDP course offered in the senior year has been the final engineering design course of the curriculum, and has been designated for summative enhancement of the expected total of gradually accumulated knowledge and skill sets of the senior-year students just before graduation. Hence, CEV492/E Graduation Design Project (GDP) offered by the ITU EEUP is the final check-point and tool/platform for the senior students and the faculty members for excelling the environmental engineering design experiences together. In other words, the PBL-driven "CEV492/E Graduation Design Project" compulsory course offered at the senior year was the "culminating major engineering design experience that 1) incorporates appropriate engineering standards and multiple constraints and 2) is based on the knowledge and skills acquired in earlier course work," as described by the ABET EAC in criteria valid for the accreditation review cycles before 2019–20 fall term, for the most recent/current ones in effect as of the 2019–20 fall term, and as the final item of the recently updated "Criterion 5. Curriculum" (ABET EAC 2018, 2022).

In addition, this senior-year engineering design compulsory course had a significant role in overall assessment of the level of attainment of several SOs at course and program levels for the ITU EEUP (Dulekgurgen et al. 2016, 2018a, 2022, Yangin-Gomec 2016, 2018).

a. *GDP rubric:* In order to measure student performance in the GDP course and to determine the effectiveness of the education in developing and/or enhancing the focused student abilities and learning attributes, a detailed and comprehensive grading rubric was created, revised, and used. Briefly, the rubric was related to the following main sections:
   - content quality and technicalities,
   - process and system design,
   - cost analysis,
   - time and project management, and
   - environmental management considerations (i.e., risk analysis, sustainability, uncertainty, life-cycle analysis, environmental impact issues).

b. *SO rubrics:* Because a genuine picture of the relevance, accuracy, and utility of selected assessment tools was essential, a remedial action – namely the PI-based assessment – was implemented in the A&E of the GDP course in order to overcome these deficiencies (Dulekgurgen et al. 2018a; Yangin-Gomec 2016). Accordingly, the instructors were asked to assess students' performances at individual PI-specificity levels and to use the detailed analytic rubrics specifically designed for each SO mapped to the GDP course.

c. *Internally developed outcome-based assessment exams (OBEx):* Senior students were given a technical exam at the end of the last semester to help assess the level of attainment of the SOs addressed by the Graduation Design Project course.

d. *Graduate Design Project presentations:* Senior students prepared oral presentations and presented their works in the *Graduation Design Project* course at the end of the semester and their team and individual performances were assessed by internal and external evaluators using the rubric designed for assessing oral presentation/communication competencies. Since rubrics are very practical and beneficial for both learning and evaluation (Chan and Ho 2019; Thambyah 2011), the rubric designed by the GDP coordination committee specifically for this course was utilized successfully, which allowed the evaluators to assign scores for achievements using the quantifiable metrics based on the extent to which the SOs were successfully attained.

In addition to the direct assessment tools and metrics listed above, some indirect assessment tools, particularly some carefully designed and selected surveys, were also used by the ITU EEUP in the SO A&E process. Note that survey questionnaires are typically used to collect feedback from the related participants within engineering education contexts, and response and completeness rates were key indicators of the quality of the data gathering method (Nicolaou and Atkinson 2019). Accordingly, the following specifically designed surveys were utilized as the main indirect assessment tools by the ITU EEUP in the SO A&E process:

6.2.6 Senior Exit Surveys

A specifically designed survey was given to all graduating seniors at the end of the semester before graduation to assess their perception of their level of attainment of the SOs. The use of senior exit surveys in assessment processes are fairly common among programs (Li and Simonson 2016; Myers et al. 2011), and the researchers have strongly agreed that undergraduate students' experiences and evaluations should certainly be integrated into the assessment of program quality in higher education (Guo 2018; Yin and Ke 2017).

6.2.7 Internship Surveys

The students of the ITU EEUP needed to perform 60 days of internship in order to improve their practical skills and theoretical knowledge. Surveys specifically designed for internship students and their employers were conducted at the end of their internship. The completed surveys were submitted to the internship committee of the department along with the internship reports.

Note that several researchers pointed to the fact that employers have concerns about the inconsistency between the needs of industry and the skill sets of engineering graduates (Volkwein et al. 2007). Accordingly, the internship surveys used by the ITU EEUP were of particular importance, not only in providing valuable data for assessment of the 11 SOs (and particularly, for SO4-multidisciplinary team work) but also in communicating the comments and reflections of the employers on that subject matter, being the primary external constituents of the program.

In the surveys in general, a 5-to-1 Likert scale (fully, mostly, minimally achieved, not achieved, unable to judge, respectively) was available for the participants to show how they considered they had achieved each of the SOs described in the related CCF (i.e., CCF being a document designed for each course and containing information about the course content, learning objectives, resources, assessment criteria, weekly schedule and the matrix mapping the course to the ABET EAC's legacy 11 SOs, etc.).

In addition to the above mentioned main direct and indirect tools, the following other direct/indirect assessment tools were considered for the assessment of level of achievement of the SOs:

Faculty survey: A survey specifically designed for faculty members that also questioned the student advisory system at EEUP. The survey helped review the PEOs and assess the SOs, the student advisory system and some specific issues related to the EEUP. Faculty surveys are addressed as an effective tool to assess and improve educational quality (Chugai 2015).

Advisory board meetings: Advisory board members were holding meetings annually in order to engage in relevant discussions regarding the level of attainment of SOs, as well as to review the PEOs of the program and discuss different career and employment issues. Information from those discussions contributed meaningfully to keeping the program's PEOs up to date in relation to the needs of the profession and the industry, to achievement of the SOs, as well as to continuous improvement of the program.

Student portfolios: Student portfolios are considered an effective tool to engage students to assess their own competencies as they progress in higher education. In this way, self-regulated learning could be provided, which is the most important achievement in education, as its logic is the stimulation for independence and self-monitoring in the learning process at different educational stages (Lam 2014). In this context, the senior-year students who teamed up in preparing their GDP reports were invited to prepare their own student portfolios, and the student team, which responded positively to that call, prepared their sample portfolio where they provided their own version of the SOs-vs-courses matrix (curriculum mapping) and their selected course works as their performance sample documents for assessing the achievement of each student outcome. Details regarding that sample student (team) portfolio and the reflections of those senior-year students are reported in details elsewhere (Dulekgurgen et al. 2022).

Student activities: Activities of student clubs such as workshops, technical site visits, and social responsibility projects were additional valuable inputs to evaluate the related SOs. In this way, undergraduate students' self-evaluation of their capacities/competencies could be supported in a holistic way across disciplines and across universities (Chan and Luk 2021).

All the above mentioned surveys were revised periodically and then were made available to the target groups through the ITU VETI –ITU's online data collection and statistics system (www.veti.itu.edu.tr), which is a web-based application system developed by the ITU IT department (www.bidb.itu.edu.tr), facilitating fast and easy collection of the necessary feedback. In this way more reliable and considerable data were gathered and fed into the continuous improvement process of the ITU EEUP.

The flowchart showing the processes used for continuous improvement of the environmental engineering undergraduate program is presented in Figure 6.2.

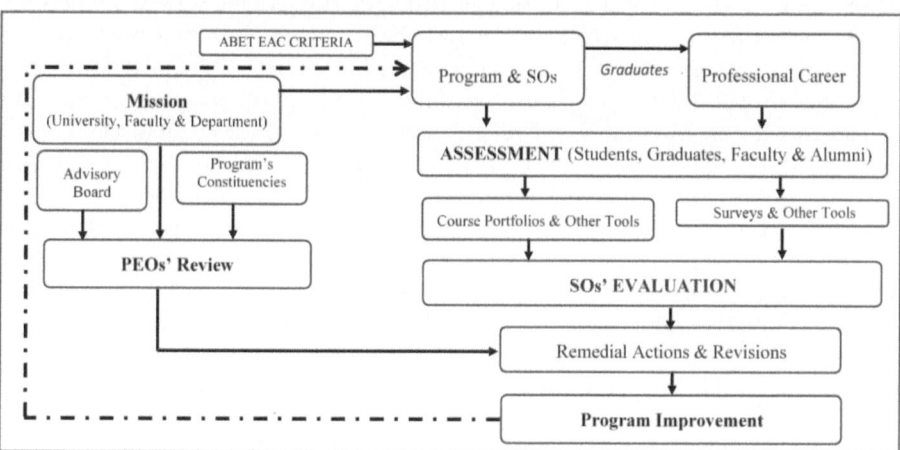

FIGURE 6.2 Review, assessment, evaluation and continuous improvement processes of ITU EEUP.

6.3 RESULTS AND DISCUSSION

6.3.1 STUDENT OUTCOMES ASSESSMENT PROCESS SPECIFIC TO THE GRADUATION DESIGN PROJECT COURSE

Assigning appropriate assessment weight ratios (%) to the direct and indirect assessment tools is considered a useful and meaningful approach with regard to good assessment practice. Accordingly, the assessment weight ratios (%) assigned to the direct and indirect assessment tools used by the ITU EEUP were discussed and determined in the joint meetings held with the participation of related departmental committees. As an example, contribution of the tools to overall SO assessment was determined as (i) 70% from direct assessment tools (i.e., courses) and (ii) 30% from indirect assessment tools (i.e., surveys, student activities).

Another important parameter in measuring the level of attainment of the defined SOs is the threshold level, due to the fact that in case the assessment results remain below the threshold, the required remedial actions should be implemented immediately so to improve the process and performances and reach at least the defined threshold. It is important to note that the attainment goal is not prescribed by the accreditation agency, and it is typically adapted depending on the needs and requirements of the program and may vary across undergraduate programs (Shafi et al. 2019). In this context, the starting value of the threshold for the level of attainment of the SOs adapted by the ITU EEUP was a relatively lower one: ≥50% of students' performances to be at/above satisfactory category (satisfactory+outstanding). Then in time, the thresholds were increased gradually (i.e., ≥70%), in compliance with the dynamic assessment and evaluation planning and the continuous improvement of the program.

Moreover, curriculum mapping – a method for relating the educational activities and the learning/student outcomes, together with the means of assessing to which extent those outcomes are being attained, including all four years of an undergraduate program – is considered a valuable and fundamental tool for SO A&E at program level (Dulekgurgen et al. 2022; UDel CTAL 2020). When given in a matrix format, standardization and visualization of the relations between the SOs and the A&E process/tools might be better provided. In addition, curriculum mapping for SO A&E would include all three levels of relative contribution of the compulsory courses offered by faculty members (namely [level-1]: Introduced; [level-2]: Reinforced; [level-3]: Emphasized; Assessed & Evaluated).

A brief example of the SO A&E scheme used by the ITU EEUP is provided below in Table 6.1, where the assessment tools and their relative contributions to the A&E are presented particularly for SO3, the latter being the student outcome most relevant to the CEV492/E GDP course: i.e., the senior-year GDP course of the program was mapped to fully satisfy the SO3-design meeting needs within realistic constraints; in other words, with the highest level of relative contribution (Emphasized; Assessed & Evaluated) in the assessment process. Similar brief assessment schemes were structured and used for assessing the attainment levels of the other SOs as well.

SOs A&E process is an effort-intensive process, requiring commitment and demanding investment of individual and collective efforts and time. On the other

TABLE 6.1

A Brief Example of SO Assessment Scheme: Assessment Methods/Tools and Thresholds Defined for A&E of SO3 at ITU EEUP

Student Outcome	Assessment Methods/Tools [Assessment Weights in %]	Time of Data Collection	Threshold
SO-3 – An ability to design a system, component, or process to meet desired needs within realistic constraints such as economic, environmental, social, political, ethical, health and safety, manufacturability, and sustainability[a]	Courses [70] Other tools [30]	2014–15 Fall–Spring 2015–16 Fall–Spring 2016–17 Fall–Spring	50% 60% 70%

[a] ABET EAC's legacy 11 SOs were in effect and adapted by the ITU EEUP at the time of this study.

hand, it is possible to optimize the process by adapting an appropriate frequency for data collection, assessment, evaluation, and improvement. When preparing for the re-accreditation cycle, the ITU EEUP performed assessment and evaluation of the ABET EAC's legacy 11 SOs on an annual basis for three consecutive academic years. However, it was decided to revise the data collection frequency and optimize the process. Accordingly, a revised/improved timetable for the SO A&E process was structured and offered to be used in the following years for the next comprehensive review cycle for re-accreditation (Table 6.2). While structuring the revised timetable, it was decided to adapt a three-year review period during which maximum of three or four SOs were to be assessed every year, followed by evaluation and recommendations in the second year, and implementation of the changes in the third year.

An example for the recommendations and implementations/remedial actions related to the continuous improvement process in the ITU EEUP is provided in

TABLE 6.2

SOs A&E Process Timetable[a]

Academic Years	SO1	SO2	SO3	SO4	SO5	SO6	SO7	SO8	SO9	SO10	SO11
1st year	A							A	A	A	
2nd year	E&R	A		A	A			E&R	E&R	E&R	A
3rd year	I	E&R	A	E&R	E&R	A	A	I	I	I	E&R
4th year	R	I	E&R	I	I	E&R	E&R	R	R	R	I
5th year		R	I	R	R	I	I				R
6th year			R			R	R				

A → Assessment; E&R → Evaluation and Recommendations; I → Implementations; R → Re-assessment

[a] ABET EAC's legacy 11 SOs were in effect and adapted by the ITU EEUP at the time of this study

TABLE 6.3

Examples for the Recommendations and Implementations/Remedial Actions Related to the Continuous Improvement for the Third Year of A&E for Some SOs and Specific to the GDP Course

SOs	Recommendations	Implementation/Action
SO1	• The assessment tools of this rubric should be revised in the GDP course.	Recommendations planned to be implemented in following academic year and students will be reassessed MODERATE REVISION
SO3	• The content of the technical drawing course is suggested to be changed/improved for better level of attainment in design drawings of the GDP course.	Recommendations planned to be implemented in following academic year and students will be reassessed MAJOR REVISION
SO8	• The revised version of GDP course ("Environmental Engineering Design Project-I" and "Environmental Engineering Design Project-II") is recommended to be offered	Recommendations planned to be implemented in following academic year and students will be reassessed MAJOR REVISION
SO10	• This SO can be assessed in the new course entitled "Environmental Engineering Design Project-II" after integration to the curriculum.	Recommendations planned to be implemented in following academic year and SO10 will be reassessed MAJOR REVISION

Table 6.3. In this SO A&E process, the term "minor revision (remedial action)" requires informative decisions to be communicated by the department curriculum development committee (DCDC) to course coordinators/instructors. The term "moderate revision (remedial action)" requires communication of the decisions from DCDC to relevant coordination commissions (i.e., GDP coordination, faculty admins, etc.). The term "major revision (remedial action)" requires communication of the decisions from DCDC to the department head and then opening them for discussion and approval by the faculty members: ITU EED academic board (DAB). In all of those SO A&E processes, assessment of the level of achievement of all 11 SOs were performed at PI-breakdown details, during which formative and summative data were collected and specifically designed analytical rubrics were used as the main direct assessment metrics. Benefits of implementing this comprehensive assessment approach are summarized in the following sections. The SO A&E results obtained through the senior-year capstone design project course, the CEV492E GDP, are discussed below.

In the GDP course, the A&E system included detailed assessment of the students' abilities and attributes corresponding to each specific PI of each SO, assessed in accordance with four performance vectors (i.e., unsatisfactory: 0–24%; developing: 25–49%; satisfactory: 50–74%; outstanding: 75–100%) and specific performance descriptors comprising the detailed analytic rubrics specifically designed for each SO. The recently recommended PI-breakdown–based assessment facilitated attainment of more realistic and meaningful results. Performance indicator (PI)–based

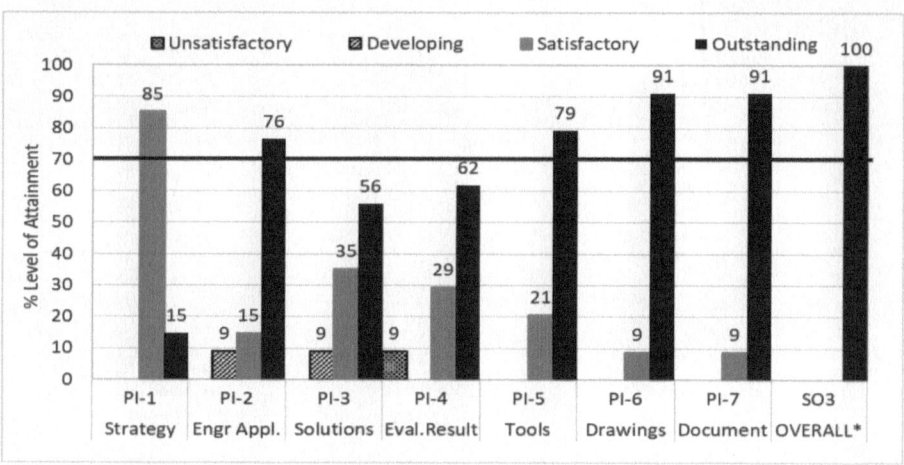

FIGURE 6.3 Comparison of overall (at or above "satisfactory") and PI-based (with performance vector details) assessment of level of attainment of SO3 by GDP in Spring 2018-19. Horizontal line shows the designated threshold.

assessment results were found to give more insight to understand the outcome attainment performance of the students at PI-specific level and to propose appropriate remedial actions when needed. Moreover, the PI-breakdown helped to determine the specific performance indicators at which students' abilities might be improved further (e.g., PI-3 and PI-4 of SO3) (Figure 6.3). In Table 6.4, the results of percentages of students achieving unsatisfactory/developing/satisfactory/outstanding levels regarding SO3 are presented with the assessment tools used at the 2018–19 spring term. According to this, informative results of the comprehensive PI-breakdown–based SO A&E for SO3 are given in for the 2018–19 spring term indicating the percentages above the category of satisfactory and percentages of all categories meeting

TABLE 6.4

Percent Meeting SO3ᵃ and the Tools Used during Evaluation

Term (Number of Students)	Percentages (%) of the Categories Meeting SO3[b]				Tool Used
	Outstanding	Satisfactory	Developing	Unsatisfactory	
2018–2019 spring (34)	91	9	0	0	Internally Developed Specific Rubric (50%) + PI-Based SO3 Rubric (50%)

[a] SO3 – an ability to design a system, component, or process to meet desired needs within realistic constraints such as economic, environmental, social, political, ethical, health and safety, manufacturability, and sustainability,

[b] Unsatisfactory: 0–24; developing: 25–49; satisfactory: 50–74; outstanding: 75–100

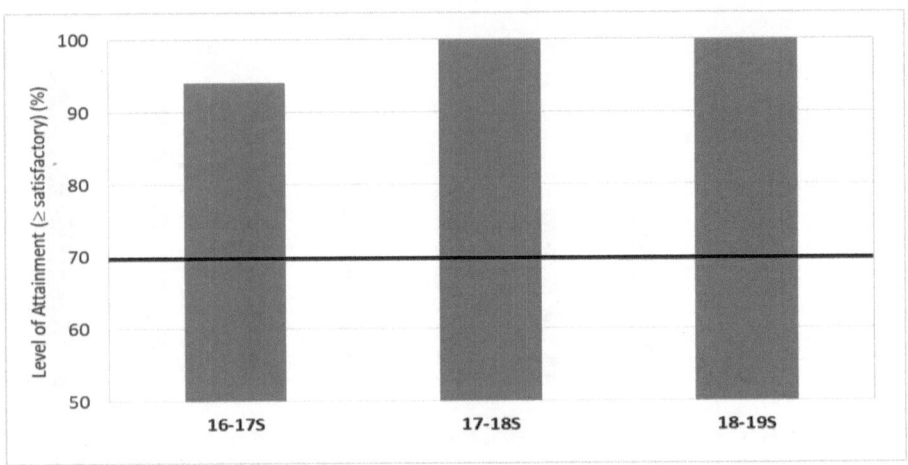

FIGURE 6.4 Overall assessment of level of attainment (at or above "satisfactory") of SO3 by the GDP in three consecutive spring semesters between the 2016–17 and 2018–19 academic years. Horizontal line shows the set threshold.

this SO. As seen in Figure 6.3, it was possible to identify at which particular PI the undergraduates seemed to be performing below expectations. Hence, application of the PI-breakdown–based assessment in the SO A&E process has been strongly recommended.

Comparative evaluation of the A&E results in three consecutive spring semesters between the 2016–17 and 2018–19 academic years showed that SO3 – "an ability to design a system, component, or process to meet desired needs within realistic constraints such as economic, environmental, social, political, ethical, health and safety, manufacturability, and sustainability" – was attained with the GDP course at a level above the set threshold (Figure 6.4). It was considered that use of the new assessment tools and implementation of the PI-breakdown–based assessment approach helped improve the level of attainment as well as increase student performances.

In addition to overall assessments for all SOs, performance-indicator (PI)–based assessments should be also available for the SOs. Compared to the overall assessment approach, the strength and advantage of the PI-based assessment is that it provides further information on the students' skills/abilities and especially about their weaknesses/deficiencies in the fields addressed by the specific PI. This helps in better identifying and understanding the assessment results and provides an opportunity for a more comprehensive evaluation of the A&E process. Through the PI-based assessment results, more insight became available to understand the outcomes-related performances of the students and to propose appropriate remedial actions. It was also considered that the improvements recorded in the level of attainment of the SOs through the GDP course might be not only because of increase in students' performances but also due to application of the aforementioned PI-breakdown–based assessment approach with the newly added assessment tools (Dulekgurgen et al. 2018a).

6.4   CONCLUSION

Higher education programs are required to generate, compile and provide proofs of program assessment practices and continuous improvement efforts so to comply with the accreditation criteria and standards. In return, those intense and continuous internal efforts serve for improvement of the program and the educational activities. Therefore, quality assurance and accreditation actions significantly contribute to the educational quality. Also, assessment processes motivate the students in directing their efforts toward effective learning and provide them appropriate feedback.

With regard to quality assurance and accreditation in higher education, the most important aspects are (i) the process for assessing and evaluating the attainment levels of the SOs, and (ii) using the information derived from that process in making decisions for improving the program, which (iii) form the basis of the continuous improvement activities. In addition, programs seeking accreditation or re-accreditation have the opportunity to examine their running processes in details, revise and optimize those, revisit their past experiences and benefit from them as lessons-learned for current and future improvement, recommend action plans for improvement and put these into action, all while preparing for comprehensive accreditation reviews requiring compliance with the quality standards specified by the accrediting agencies. In this respect, educational quality assurance and accreditation efforts should put the utmost emphasis on successful realization of the student outcomes in a particular undergraduate/graduate program, and particularly in engineering undergraduate programs in general, that have started to be aligned with the outcomes conditioned by their respective criteria.

Last but not least, results of the SO A&E process run by the ITU EEUP reflect that the recently recommended and implemented PI-breakdown–based SO assessment facilitates attainment of more realistic, meaningful, and easily assessable assessment results. Finally, as the PI-breakdown– based SO assessment approach requires advance planning and detailed documentation of the assessment tools selected for assessing each one of the PIs of every SOs, implementation of this approach enables elaboration of the SO assessment process, including, for example, checking the relevance and accuracy of the preferred assessment tools. Those provide a comprehensive picture of the strengths and weakness at various levels, enabling determination of the points of interest for further improving the educational activities and eventually the program.

ACKNOWLEDGEMENT

We would like to acknowledge all the faculty members of the ITU EED who took part in preparation of the self-study report.

We would like to express our deepest condolences for our beloved friend and colleague Assoc. Prof. Didem Okutman Tas, whose exquisite work in this study is deeply appreciated. Without her precious contribution, this study would not have been possible. Her absence is a grave loss for our academia but her dedication to science and her students will continue to inspire us all. May she rest in peace.

REFERENCES

ABET (2022). Official website http://www.abet.org/; Find an ABET-Accredited Program. https://amspub.abet.org/aps/name-search?searchType=institution&keyword= Istanbul%20Technical%20University. Last accessed 12 Feb 2022.

ABET EAC, Baltimore, MD, USA. "Criteria for Accrediting Engineering Programs, 2019 – 2020" (2018). https://www.abet.org/accreditation/accreditation-criteria/criteria-for-accrediting-engineering-programs-2019-2020/#GC3 First accessed 11 December 2018, Last accessed 07 July 2020.

ABET EAC, Baltimore, MD, USA. "Criteria for Accrediting Engineering Programs, 2022 – 2023" (2022). Last accessed 12 Feb 2022.

Akduman, I., Özkale, L., Ekinci, E. (2001). Accreditation in Turkish universities, European Journal of Engineering Education, 26(3), 231–239. DOI: 10.1080/03043790110053374

Aoudia, M., Abu-Alqahsi, D. A.-D. (2015). Curriculum redesign process for an industrial engineering program seeking ABET accreditation. International Journal of Engineering Pedagogy (iJEP), 5(3), 45–52. https://doi.org/10.3991/ijep.v5i3.4670

Asif, M., Awan, M.U., Khan, M.K., Ahmad, N. (2013). A model for total quality management in higher education. Quality & Quantity, 47(4), 1883–1904. 10.1007/s11135-011-9632-9

Barrett, B., Fernandez, F., Gonzalez, E. M. (2020). Why universities voluntarily pursue US accreditation: The case of Mexico. Higher Education, 79, 619–635. https://doi.org/10.1007/s10734-019-00427-y

Biggs, J. B., Tang, C. S. (2011). Teaching for quality learning at university: What the student does. Philadelphia, PA: McGraw-Hill/Society for Research into Higher Education.

Chan, Z., Ho, S. (2019). Good and bad practices in rubrics: The perspectives of students and educators. Assessment & Evaluation in Higher Education, 44(4), 533–545. DOI: 10.1080/02602938.2018.1522528

Chan, C K.Y., Luk, L.Y.Y. (2021). Development and validation of an instrument measuring undergraduate students' perceived holistic competencies. Assessment & Evaluation in Higher Education, 46(3), 467–482. DOI: 10.1080/02602938.2020.1784392

Chugai, O.Y. (2015). Defining, assessing and improving teacher quality in the USA. Advanced Education, 4, 60–65. https://doi.org/10.20535/2410-8286.51350

Das, S., Biswas, S., Biswas, A. (2017). Impact of advance pedagogy in engineering and outcome based education. Communications on Applied Electronics 6(8), 22–27. DOI: 10.5120/cae2017652537

Dayananda, P., Latte, M. V., Raisinghani, M. S. Sowmyarani C.N. (2020). New approach for target setting mechanism of course outcomes in higher education accreditation. Journal of Economic and Administrative Sciences, 37(1), 79–89. http://dx.doi.org/10.1108/JEAS-03-2020-0024

Dulekgurgen, E., Okutman, D., Yangin-Gomec, C (2022). Curriculum mapping for attainment of student outcomes from the perspectives of senior-year students and instructors. In Kaushik Kumar (ed) Engineering Pedagogy Towards Outcome-Based Education, Chapter 10. CRC Press, Taylor & Francis Group, LLC, Boca Raton, FL. ISBN 9780367537432.

Dulekgurgen, E., Ozgun, O.K., Yuksek, G., Pasaoglu, M.E., Unalan, C., Bicer, O.B., Cetinkaya, Z., Isik, I., Oner, B.E. (2016). A final touch for the environmental engineering students at the onset of their profession: Senior-year graduation design project – Case Study for 2014–2015. International Journal of Engineering Pedagogy 6(2), 23–29. https://doi.org/10.3991/ijep.v6i2.5375.

Dulekgurgen, E., Taptik, Y., Aydin, A.F. (2018b). ABET EAC accreditation processes at ITU and educational quality evaluations – from tradition to future. ITU Journal-Special Issue: Quality in Higher Education: İTÜ Vakfı Dergisi – İstanbul Teknik Üniversitesi Vakfı Yayını, Jan-March 2018, (79), 42–47 (in Turkish).

Dulekgurgen, E., Yangin-Gomec, C., Ozgun, O. K., Aydin, B., Guven, H. (2018a). Follow-up on assessment of student outcomes by senior-year design project and continuing to improve by performance indicator breakdown-based assessment. International Journal of Engineering Pedagogy, 8(5), 19–28. https://doi.org/10.3991/ijep.v8i5.8148

Dulekgurgen, E. (2018c). Quality assurance of engineering education and student outcome-based assessment and evaluation: The ITU – ABET EAC example. *Presentation of the ITU Delegation (ITU IAC, ODoS, CE3) in the YOKAK (THEQC) Conference on "Establishment of an Internal Quality Assurance System in Higher Education Institutions and Quality Processes in Education."* 27–28 Nov 2018, YOK Conference Hall, Bilkent, Ankara, Turkey (in Turkish). https://api.yokak.gov.tr/Storage/AnnouncementFiles/29-11-2018/34/Ebru%20Dulekgurgen.pdf. Last accessed 12 Feb 2022.

Dulekgurgen, E., Taptik, I.Y, Aydın, A.F., Hosni, M. (2019b). Power of continuous-knowledge sharing in preparing for successful accreditation reviews. *In BUA2019 Online Proceedings (Full Papers/Working Papers) – 5th Annual Conference of the Balkan Universities Association*, 16–18 April, 2019, Thessaloniki, Greece. http://bua2019.web.auth.gr/wp-content/uploads/2019/04/BUA2019_WorkingPapers.pdf; pp.146–152.

Dulekgurgen, E., Taptik, I.Y, Aydın, A.F., Karaca, M. (2019a). International accreditation as a means of improving engineering education: the ITU – ABET experience. *In BUA2019 Online Proceedings (Full Papers/Working Papers) – 5th Annual Conference of the Balkan Universities Association*, 16–18 April, 2019, Thessaloniki, Greece. http://bua2019.web.auth.gr/wp-content/uploads/2019/04/BUA2019_WorkingPapers.pdf; pp.134–140.

Dym, C. L., Agogino, A. M., Eris, O., Frey, D. D., Leifer, L. J. (2005). Engineering design thinking, teaching, and learning. Journal of Engineering Education, 34(1), 103–120. https://doi.org/10.1002/j.2168-9830.2005.tb00832.x

EC2000 (*aka ABET Engineering Criteria 2000*). (1995). Criteria for accrediting programs in engineering in the United States, ABET, Inc., Baltimore, MD, December 1995.

ElAli, T. (2017). Innovation in engineering education: A proposed ABET course outcomes assessment portfolio. International Journal of Engineering and Advanced Research Technology, 3(7), 14–18. http://dx.doi.org/10.20431/2349-4050.0503004

Felder, R. M., Brent, R. (2003). Designing and teaching courses to satisfy the ABET Engineering Criteria. Journal of Engineering Education, 92(1), 7–25. https://doi.org/10.1002/j.2168-9830.2003.tb00734.x

Fowler, D., Froyd, J.E., Layne, J. (2010). Curriculum redesign: Concurrently addressing content mastery and development of cognitive abilities. 2010 IEEE Frontiers in Education Conference (FIE), T1H-1-T1H-5.

Frank, A., Kurth, D., Mironowicz, I. (2012). Accreditation and quality assurance for professional degree programmes: Comparing approaches in three European countries. Quality in Higher Education, 18(1), 75–95. DOI: 10.1080/13538322.2012.669910

Frazer, M. (1992). Quality assurance in higher education. In Alma Craft (ed.) *Quality Assurance In Higher Education*, pp. 9–26, The Falmer Press, Taylor & Francis Inc., ISBN 0-203-26766-4.

Gerritsen-van Leeuwenkamp, K. J., Joosten-ten Brinke, D., Kester, L. (2019). Students' perceptions of assessment quality related to their learning approaches and learning outcomes. Studies in Educational Evaluation, 63, 72–82. https://doi.org/10.1016/j.stueduc.2019.07.005.

Guo, J. (2018). Building bridges to student learning: Perceptions of the learning environment, engagement, and learning outcomes among Chinese undergraduates. Studies in Educational Evaluation, 59, 195–208. https://doi.org/10.1016/j.stueduc.2018.08.002.

ITU EED (2022). Istanbul Technical University, Environmental Engineering Department official accreditation info web-page: "ABET Accreditation (itu.edu.tr)" (2022). Last accessed 12 Feb 2022. https://cevre.itu.edu.tr/en/education/accreditation/abet-accreditation

Jacob, S. A., Boyter, A. (2020). It has very good intentions but it's not quite there yet: Graduates' feedback of experiential learning in an MPharm programme Part 2 (TELL Project). Studies in Educational Evaluation, 66, 100889.

Jongbloed, B., Vossensteyn, H., van Vught, F., Westerheijden D.F. (2018) Transparency in higher education: The emergence of a new perspective on higher education governance. In Curaj A., Deca L., Pricopie R. (eds.) *European Higher Education Area: The Impact of Past and Future Policies.* Springer, Cham. https://doi.org/10.1007/978-3-319-77407-7_27

Kumpas-Lenk, K., Eisenschmidt, E., Veispak, A. (2018). Does the design of learning outcomes matter from students' perspective? Studies in Educational Evaluation, 59, 179–186.

Lam, R. (2014). Promoting self-regulated learning through portfolio assessment: Testimony and recommendations. Assessment & Evaluation in Higher Education, 39(6), 699–714. http://dx.doi.org/10.1080/02602938.2013.862211

Leenknecht, M., Wijnia, L., Köhlen, M., Fryer, L., Rikers, R., Loyens, S. (2021). Formative assessment as practice: The role of students' motivation. Assessment & Evaluation in Higher Education, 46(2), 236–255. DOI: 10.1080/02602938.2020.1765228

Li, I., Simonson R. D. (2016). The value of a redesigned program and capstone course in economics. International Review of Economics Education, 22, 48–58. DOI: 10.1016/j. iree.2016.05.001

Mintz, K., Tal, T. (2014). Sustainability in higher education courses: Multiple learning outcomes. Studies in Educational Evaluation, 41, 113–123. https://doi.org/10.1016/j. stueduc.2013.11.003

Morell, L. (2010). Engineering education in the 21st century: Roles, opportunities and challenges. International Journal of Technology and Engineering Education, 7(2), 1–10. http://ijtee.org/ijtee/system/db/pdf/72.pdf#page=5

MUDEK (2022). Official web-site of Association for Evaluation and Accreditation of Engineering Programs. https://www.mudek.org.tr/en/ana/ilk.shtm. Last accessed 12 Feb 2022.

Myers, S. C., Nelson, M. A., Stratton, R. W. (2011). Assessment of the undergraduate economics major: A national survey. The Journal of Economic Education, 42(2), 195–199. DOI: 10.1080/00220485.2011.555722

Nicolaou, M., Atkinson, M. (2019). Do student and survey characteristics affect the quality of UK undergraduate medical education course evaluation? A systematic review of the literature. Studies in Educational Evaluation, 62, 92–103. https://doi.org/10.1016/j. stueduc.2019.04.011

Orsmond, P., Merry, S., Sheffield, D. (2006). A quantitative and qualitative study of changes in the use of learning outcomes and distractions by students and tutors during a biology poster assessment. Studies in Educational Evaluation, 32(3), 262–287. https://doi. org/10.1016/j.stueduc.2006.08.005

Öz, H. H. (2005). Accreditation processes in Turkish higher education. Higher Education in Europe, 30(3-4), 335–344. DOI: 10.1080/03797720600625960

Passow, H. J., Passow, C. H. (2017). What competencies should undergraduate engineering programs emphasize? A systematic review. Journal of Engineering Education, 106(3), 475–526. DOI: 10.1002/jee.20171

Prados, J. W., Peterson, G. D., Lattuca, L. R. (2005). Quality assurance of engineering education through accreditation: The impact of Engineering Criteria 2000 and its global influence. Journal of Engineering Education, 94(1), 165–184. https://doi.org/10.1002/ j.2168-9830.2005.tb00836.x

Ramírez, G. B., Luu, D. H. (2018). A qualitative exploration of motivations and challenges for implementing US accreditation in three Canadian universities. Studies in Higher Education, 43(6), 989–1001. DOI: 10.1080/03075079.2016.1203891

Ryan, T. (2015). Quality assurance in higher education: A review of literature. Higher Learning Research Communications, 5(4). DOI:10.18870/hlrc.v5i4.257

Saribeyli, F. R. (2018). Theoretical and practical aspects of student self-assessment. The Education and Science Journal, 20(6), 183–194. DOI: 10.17853/1994-5639-2018-6-183-194

Segers, M., Dochy, F., Gijbels, D., Struyven, K. (2009). Changing insights in the domain of assessment in higher education: Novel assessments and their pre-, post- and pure effects on student learning. In D.M. McInerney, G.T.L. Brown, G.A.D. Liem (eds.) Student Perspectives on Assessment: What Students can Tell us about Assessment for Learning. Information Age Publishing, INC, Charlotte, NC, pp. 297–319.

Shafi, A., Saeed, S., Bamarouf, Y. A., Iqbal, S. Z., Min-Allah, N., Alqahtani, M. A. (2019). Student outcomes assessment methodology for ABET accreditation: A case study of computer science and computer information systems programs. IEEE-ACCESS 7: 13653–13667. DOI: 10.1109/ACCESS.2019.2894066.

Sin, C., Orlanda, T., Amaral, A. (2017). The impact of programme accreditation on Portuguese higher education provision. Assessment & Evaluation in Higher Education, 42(6), 860–871. DOI: 10.1080/02602938.2016.1203860

Strijbos, J., Engels, N., Struyven, K. (2015). Criteria and standards of generic competences at bachelor degree level: A review study. Educational Research Review, 14, 18–32. https://doi.org/10.1016/j.edurev.2015.01.001

Struyven, K., Blieck, Y., De Roeck, V. (2014). The electronic portfolio as a tool to develop and assess pre-service student teaching competences: Challenges for quality. Studies in Educational Evaluation, 43, 40–54. https://doi.org/10.1016/j.stueduc.2014.06.001

Stura, I., Gentile, T., Migliaretti, G., Vesce, E. (2019). Accreditation in higher education: Does disciplinary matter? Studies in Educational Evaluation, 63, 41–47. DOI:10.1016/J.STUEDUC.2019.07.004

Thambyah, A. (2011). On the design of learning outcomes for the undergraduate engineer's final year project. European Journal of Engineering Education, 36(1), 35–46, DOI: 10.1080/03043797.2010.528559

UDel CTAL: University of Delaware – Center for Teaching and Assessment of Learning official web-site, "Curriculum Mapping" (2020). https://ctal.udel.edu/resources-2/mapping/ Accessed 07 July 2020

Ulker, N., Bakioglu, A. (2019). An international research on the influence of accreditation on academic quality. Studies in Higher Education, 44(9), 1507–1518, DOI: 10.1080/03075079.2018.1445986

van der Vleuten, C., Sluijsmans, D., Joosten-ten Brinke, D. (2017). Competence assessment as learner support in education. In Mulder M. (eds) Competence-based Vocational and Professional Education. Technical and Vocational Education and Training: Issues, Concerns and Prospects, vol. 23. Springer, Cham. https://doi.org/10.1007/978-3-319-41713-4_28

Volkwein, J.F., Lattuca, L.R., Harper, B.J., Domingo, R.J. (2007). Measuring the impact of professional accreditation on student experiences and learning outcomes. Research in Higher Education, 48, 251–282. https://doi.org/10.1007/s11162-006-9039-y

Vu, T. T., Dall'Alba, G. (2007). Students' experience of peer assessment in a professional course. Assessment & Evaluation in Higher Education, 32(5), 541–556. DOI: 10.1080/02602930601116896

Wilkerson, J. R. (2017). Navigating similarities and differences in national and international accreditation standards: A proposed approach using US agency requirements. Quality Assurance in Education, 25(2), 126–145.

Yangin-Gomec, C., Kose-Mutlu, B., Dulekgurgen, E., Ozturk, I., Tanik, A. (2016). Development and application of a rubric specific for the senior-year Graduation Design Projects for assessing learning outcomes. Iberoamerican Journal of Project Management (IJoPM), 7(1), 47–61.

Yangin-Gomec, C., Zengin, G. E., Kaynak, B., Tanik, A., Okutman Tas, D. (2018). Assessing and evaluating the student outcomes in line with continuous improvement process: A case study in environmental engineering undergraduate program. 12th International Technology, Education and Development Conference (INTED), Valencia, Spain, March 05-07, 2018, Book Series: INTED Proceedings; Pages: 99–106. DOI: 10.21125/inted.2018.1014

Yin, H., Ke, Z. (2017). Students' course experience and engagement: An attempt to bridge two lines of research on the quality of undergraduate education. Assessment and Evaluation in Higher Education, 42(7), 1145–1158. DOI: 10.1080/02602938.2016.1235679

YOKAK (2022). Official web-site of The Higher Education Quality Council of Turkey-THEQC. https://yokak.gov.tr/ Last accessed 12 Feb 2022.

7 Significance of Advanced Pedagogies, Approaches, and Frameworks in Outcome-Based Education

T. Nivethitha and P. K. Poonguzhali
Hindusthan College of Engineering and Technology
Coimbatore, Tamil Nadu, India

"Engineering is the profession in which a knowledge of mathematics and natural sciences gained by study, experience, and practice is applied with judgment to develop ways to utilize, economically, the materials and forces of nature for the benefit of mankind."

ABET (Accreditation Board for Engineering and Technology)

CONTENTS

DOI: 10.1201/9781003083160-10

7.1 INTRODUCTION

To begin, we confess a bias in favor of philosophy by starting with philosophy rather than engineering. Our approach could be enhanced by considering engineering philosophy from the perspectives of a broad range of engineering disciplines and other disciplines – possibly in connection to their mixture of ancient or cultural forms [1]. Still, our preference is toward engineering philosophy for philosophers rather than engineers, though we hope that some engineers will find it useful. Subsequently, engineering and technology are rarely entirely and clearly differentiated. Indeed, we feel that the divide between engineering and technology [2] is a philosophical issue that could be addressed in numerous ways that are complementary.

7.2 EXPECTATIONS FROM ENGINEERING EDUCATION

Our country generates a huge number of engineers each year; however the industrial sector has expressed dissatisfaction with the shortage of engineers of the desired caliber. This is a cause for concern because the significance and excellence of engineering education in India is critical to the country's industrial prosperity. After completing their pre-university education, the majority of India's brilliant students choose to pursue engineering studies. This may be due to an increase in the stature of engineering in the middle and higher classes, as well as the dominance of engineering and information technology in the worldwide market. The rising call for technical education has led in the establishment of more engineering colleges [3], but the low quality of education has resulted in a decrease in the number of engineering institutions. Soft skills are personality traits and behaviors that will help candidates get hired and succeed in their work and they are listed below:

- Approach (sincerity, can-do, ownership/motivation)
- Commerce morals/sincerity
- Confidence/self-assurance
- Contact skills
- Common consciousness
- Essential executive skills (guidance, joint effort, point-in-time administration, etc.)
- Vital action and client overhaul (most entry level jobs necessitate one of these)
- Actions to be taken in advance of an occurrence or undesirable event

Soft skills are considered to be most important component of the above skill sets. Communication skills appear to be closely linked to a positive mindset.

In order to obtain the skill sets from engineers, industries need institutions to facilitate students in accordance with industry needs. That is why, rather than traditional education, outcome-based education (OBE) is the most important. Thank you to the Indian government for taking steps in this direction, such as the National Assessment and Accreditation Council (NAAC) and National Board of Accreditation (NBA), and for enforcing tight rules and regulations for certification and assessment.

7.3   THE IMPORTANCE OF OUTCOME-BASED EDUCATION

Whether it's in education or business, the globe is undergoing many transitions at breakneck speed. The need for T-shaped abilities or T-shaped people is exceptionally high at such a tumultuous time. The T-shape model is a metaphor for describing a person's abilities in the workplace that is used in job recruitment. The horizontal bar on the letter T signifies one's capacity to cooperate across disciplines and apply information in corresponding areas of competence, whereas the vertical bar represents one's depth of knowledge and experience in a single field [4].

Companies must improve their competitiveness through enhancing human resources in our current environment, which is replete with socio-cultural, economic, and demographic changes. Outcome-based education revolutionizes curriculums to help graduates achieve this goal by integrating particular awareness with vibrant and cross-sectional ability.

7.4   OBE VERSUS TRADITIONAL EDUCATION

Today, change is alone permanent and arrives with the necessity for educational organizations to update and familiarize themselves with new approaches or face the threat of becoming out of date. OBE is an educational representation in which core programs, pedagogy, and evaluation processes are restructured to replicate the attainment of higher-order learning rather than the accumulation of course credits. Unlike traditional learning where approach is exam-driven, in OBE Learners are assessed on an ongoing basis. It is a student-centered approach that includes real-world experience. What matters more than what or how something is taught is the knowledge, abilities, and traits that students get at the end of a program or course. Traditional education is based on standardized practices in which students gather under one roof at a specific time to be educated by a teacher. Learners interact with peers or ask faculty members questions once a lecture is over. This suggests that the effectiveness of the educational system is mostly determined by the teacher's efficacy and classmates' knowledge base. On the other hand, OBE is a system of education based on precise outcome [5]. It emphasizes the expertise that students will need when they finish their education. Activities in and out of the classroom are planned to assist students in achieving these goals.

7.5   SORTING OUT THE ADVANTAGES AND DRAWBACKS

One of the significant advantages of OBE is the feeling of precision it provides. Based on explicitly stated learning objectives, students and their parents can choose an institution, program, and course. The program educational objective (PEO), program outcome (PO), course outcome (CO), and program specific outcome (PSO) all define what students are expected to do after finishing a course or program. This precision is mirrored in the importance of schooling and sharing out in departments, wherever staff can more appropriately adapt their focus. The next arguably obvious main advantage is elasticity. OBE gives learner the freedom to pick what they would like to learn and how they desire to study it. It not only adapts to a learner's strengths and shortcomings, but it also gives them enough time to become proficient and fluent in the topic. Furthermore, the model permits students to relocate their credits and move to another university that follows the OBE syllabus. This accreditation allows organizations to be renowned, benchmarked, and easily compared to one another. As you can see, the OBE framework helps all stakeholders.

OBE, like any instructional approach, comes with its own set of difficulties. The aspect of interpretation is first and foremost. A prescriptive set of instructional design is missing from the framework. While the results are plainly shown, there is a lot of opportunity for understanding. Plus, with so much waffle, it's easy to get caught up in the technical terms instead of focusing on the implications of each one. Constructing learning outcomes can be challenging and time-consuming.

This leads us to the next disadvantage. In comparison to the arts, OBE works well with job-related education courses such as engineering and science. Subjects like nonfiction and philosophy, for example, entail a free-flowing organization. The main drawback, however, may be related to the assessment aspect. The best in OBE isn't brought out by paper-and-pencil tests. It necessitates a variety of examinations, ranging from cluster projects to interrogation. In addition, evaluating OBE in an effective environment is problematic to deal with the disadvantages; the key is to strike an equilibrium between what is anticipated and feasible.

Looking into the expectations, do you realize that the entirety of today's students intend to be entering jobs that don't exist? Technological breakthroughs, together with socio-cultural, economic, and demographic developments, are disrupting the business world and will persist in doing so. As a result, OBE will be on the cusp of a new sphere, one in which pupils will be expected to manage a constantly shifting global terrain. Here's what we can anticipate. A growing number of students are looking for new abilities to help them advance in their careers. Demand is increasing for vocational training, degrees that can be changed, and programs that are competency based. Plus, there's more. Teachers may progress from being knowledge disseminators to knowledge facilitators.

7.6   INFLUENCE OF ADVANCES IN PEDAGOGY IN ENGINEERING AND OUTCOME-BASED EDUCATION

In order to draw more young people into engineering and confirm that they are well prepared to address potential professional situations, we need to understand how successful engineers reflect and proceed when confronted with difficult situations. Making narrative thoughts and approaches into elucidation that fulfill original necessities, or adding considerable change to existing products and services is what innovation is all about. New technical and training needs in engineering education necessitate curriculum innovation, while educational practice innovation can boost student learning and staff productivity. The pedagogical approach as well as the introduction of advanced pedagogy in engineering [5] can help achieve this goal. Micro-teaching is a major component of advanced education, in which each topic is evaluated and taught using real-world philosophy. The main principle of learning outcomes is content-level investigation and

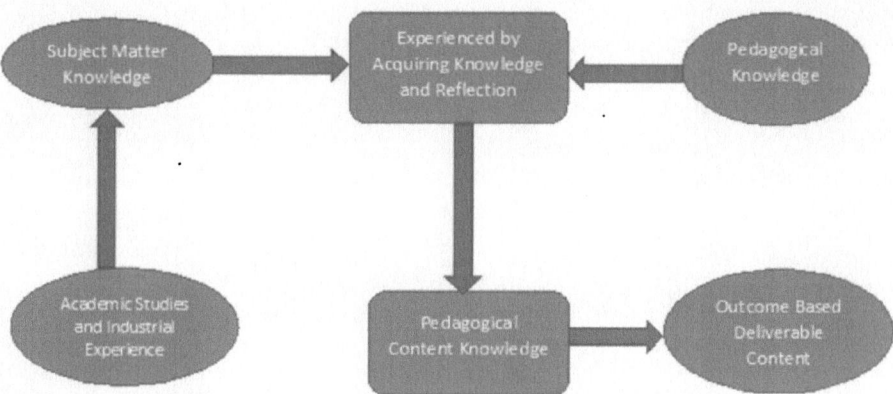

FIGURE 7.1 Structure of engineering pedagogy and OBE. (https://www.researchgate.net/figure/Key-processes-in-outcome-based-education_fig1_228692633)

application to real-world philosophy. Figure 7.1 depicts the structure of engineering pedagogy and OBE.

7.7 FRAMEWORK OF ENGINEERING EDUCATION

By retaining graduate attributes (GAs), the engineering skeleton [6] is used as a device for assessing the level in order to uphold scholastic standards, curriculum, and training practices. Figure 7.2 depicts the framework of outcome-based education. The graduate attributes are as follows:

1. An engineering knowledge base (KB): proficiency in natural sciences, engineering principles, university-level mathematics and focused engineering, knowledge relevant to the curriculum.
2. Problem investigation: the capacity to identify, create, evaluate, and solve complex engineering problems using suitable knowledge and abilities in order to achieve justified findings.
3. Investigation: the ability to examine complex problems using procedures such as proper experiments, data analysis and interpretation, and information synthesis to arrive at accurate conclusions.
4. Apparatus and engineering tool use: the capacity to design, choose, familiarize oneself and expand suitable and relevant methodologies, possessions, and current engineering tackle diversity of engineering operations – from easy to composite, while being aware of their limitations.
5. Teamwork: the capacity to operate efficiently as a team associate and manager, ideally in a multidisciplinary environment.
6. Communication skills: the capacity to convey complicated technical concepts both within the profession and to the general public. Reading, writing, speaking, and listening are examples of such abilities, as they are capacity to deduce and inscribe good information, as well as the ability to offer and successfully react to apparent instructions.

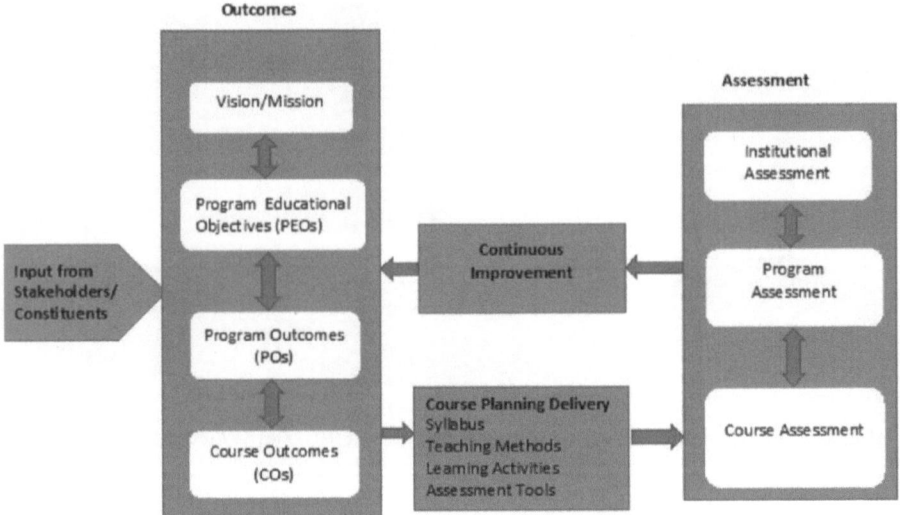

FIGURE 7.2   Outcome-based framework. (https://www.msuiit.edu.ph)

7. Professionalism: a grasp of the professional engineer's liability and obligations in society, mainly the fundamental duty of public safety and public interest protection.

8. Ethical principles of responsibility: Professional abilities such as "understanding of code of ethics, personal accountability" and "the broad education required for understanding the impact of engineering solutions on a global, economic, environmental, and societal context" are required [7].

9. Economics and project management: the capacity to integrate finances and commerce techniques into engineering practice – particularly project, risk, and change management – and to appreciate their limitations.

10. Lifetime learning: the capability to distinguish and solve one's personal learning needs in a changing world in ways that allow one to maintain proficiency and contribute to knowledge advancement.

7.8   PHILOSOPHY OF ENGINEERING

Engineering philosophy is an emerging field whose views should be considered by engineers. Philosophical thought, including replication, is significant technique to improve engineering judgment because of the empirical character of engineering practice. The differences between engineering education and practice and other areas are discussed, as well as how designers of non-prototypically designed systems deal with a substantial amount of uncertainty and how aesthetics fit into an overall sense of engineering quality [8]. The references in this paper do not represent an exhaustive list of all papers on engineering philosophy, but they are a good starting point for the reader to delve into the appropriate literature and then go as deep or as generally as desired.

7.9   DEMARCATING ENGINEERING

Engineering makes use of scientific knowledge, but it is not merely applied science, despite popular belief. Furthermore, engineering and science have different goals: science seeks information, while engineering seeks meaningful change. It seeks impartial knowledge – truth that is enduring and grounded in reality – for the sake of intellectual reflection and understanding. Engineering, on the other hand, is defined by possibility, likelihood, individuality, and compactness. Engineers count on subjective expertise and thoughts based on private and historical experience, with the purpose of deliberate action and application.

The labor of wealthy aristocrats or information seekers with affluent patrons who looked to the stars to comprehend the origins of the cosmos and life or who were intrigued to grasp the natural physical, chemical, or biological world around them gave rise to science. Engineering commenced in the trades, with the goal of creating and implementing something useful, initially for military use and then for civilian use.

Even engineering is not the same as science. The main reason for this is that engineering science was created to aid in the creation of engineering artifact. One notable difference between thermodynamics taught as physics and thermodynamics taught as engineering science is that design demands the use of control volumes, which isolate the object of interest from the rest of the world. We are concerned with the behavior of our artifact during the design stage, and we are not interested in the region beyond our design except to the extent that it has impacts on our artefact that we must consider in our scheme. As a result, the input and output points are the only points of contact with the outside world.

7.10   METHODOLOGIES

In order to improve the employability of graduates, a number of skills have to be identified for incorporation throughout the scholastic path of learners [9]. The purpose of the curriculum is to endow with undergraduate learners from all disciplines through numerous opportunities to apply basic procedural facts while also developing their personal skill sets.

To achieve these goals, an updated curriculum enhances the learning practice of students by incorporating valid and research-oriented learning approaches into a set of (mainly) current engineering and computer science degree programs. While some of these features spanned across fields, others remained focused on a particular discipline.

Engineers from many disciplines were encouraged to congregate more prolifically and allocate general ground in the fields of engineering, analytics, relevance of engineering notional knowledge, and skill expansion.

At the end of this process, five main principles were recognized as being able to lead the implementation of the revised program and distribute the following broad erudition outcomes:

- interregulation
- problem-based learning

- legitimacy
- entrenched skills
- endorsement of intrinsic motivation
- programmed structure

The endeavor to examine the curriculum resulted in an inter-departmental program that is linked in two ways. To begin, the core courses have been designed to combine conjecture and follow up in the course of student-centered activities that require students to use and apply their knowledge and skills in real-world situations. In addition, common projects, syllabi, and cross-disciplinary learning bring students from several departments together.

Various professional programs are among the subjects covered by the degrees. The changes preserved direct student access in each area, but they also established a universal configuration that allowed for standard cross-cutting action.

Design initiatives form the core strand. There are three categories of these:

- Engineering challenges: Project-based Learning (PBL) is a great structure for students to practice collaboration within teams. Student groups create solutions to a real-world predicament that incorporate all aspects of the intend process.
- Scenarios: Students participate in six discipline-based, exhaustive week-long projects during which they are not able to participate in any other classes. These are the most important prospects for graduates to link the conjecture they've learned and apply concepts from the proposal and professional skills stream into practice. They usually pursue a four-week sequence of "conventional" instruction followed by each one-week scenario.

Two new courses were established to underlie and support the abilities that students need to maximize education prospects throughout the series of design projects:

- Mathematical analysis and modeling: Both years provide an applied mathematics program that includes computer modelling. It is taught in a hybrid format that includes broad lectures supported by discipline-specific practicums. The goal is to teach fundamental mathematical principles with a high level of engineering applicability.

These new modules are supplemented with discipline-specific methodological teaching, which includes core material provided by each department and still accounts for roughly 75% of the entire degree.

7.11 PHILOSOPHIES AND PEDAGOGIES

A goal is to describe how the described educational philosophies present themselves as pedagogies in the program's curriculum parts. We define each ideology and show how it manifests itself in the individual education programs (IEP)'s learning activities.

7.12   INTERDISCIPLINARY TECHNIQUES

Proficient engineers do not work in silos based on their disciplines. It's critical for engineering students to see and understand how multiple disciplines work together to complete any contemporary engineering scheme. The suggestion to improve training and erudition across disciplines, on the other hand, is far from simple. One noticeable lane was to follow wide-ranging engineering programs, such as those presented by Oxford and Cambridge in the United Kingdom and many other parts of Europe, in which the beginning period of the degree program is general to all disciplines, with specialization happening only in the latter years. This can be accomplished by combining the teaching of courses such as mathematics, programming, and key science concepts such as fluids, thermodynamics, and so on.

Mathematics was the only important common ground that emerged from the old subjects. There was an opportunity to affix value to the students' practice by giving them a broader picture of calculations and a knowledge that it is a public tool used outside of their fields of study. An electrical engineer, for example, can benefit from understanding that the differential equations that support circuit theory can also be useful to the mechanism and constraint in a car's deferment system or the mixing of fluid flows in a chemical plant. This type of learning provides our students with a common language and culture that they recognize as underlying engineering as a whole, as well as the ability to engage in interdisciplinary projects [10].

All of these skills are necessary in project work, thus connecting the courses through projects made sense. Individual perspectives and varied perspectives are the true worth of interdisciplinary projects.

7.13   PROBLEM-BASED LEARNING

Skill strengthening is not only the goal identified by the industry. Graduate employability is the highest priority to make engineering a creative solution to societal problems. The need is to develop innovative solutions that bring together people with diverse expertise and capabilities in interdisciplinary and inclusive activity. When it comes to how these activities are structured and where they are placed in the curriculum, their impact on learner commitment and self-value is critical.

7.14   AUTHENTICITY

The provision of reliable education in engineering and cross-curricular disciplines has been a cornerstone of the IEP's concept. The education curriculum for undergraduate engineering education is supposed to serve as a basis for students to create and grow their engineering practical competence. To do this, the syllabus should provide a variety of genuine responsibilities that involve the integration of varied information and abilities.

Authentic learning is a generic model or educational approach, not a learning conjecture or a specific pedagogy [11]. We pass on to creating pedagogies in which learning objectives are aligned with real-world tasks, content, and context in a broad sense. The goal has been set for students to develop as self-indulgent and gain abilities by incorporating prior practice, personal inquiry, and awareness of environments and opportunities that will be provided.

Learning environments are those where students are divided among groups, and here their self-learning processes begins to mimic the the type of processes that engineers go through on a daily basis. The way made students acquire technical knowledge is not detached from how it will be used in the future. Rather, our students must consider, "How can I apply this technique?" or "Which modus operandi is most appropriate for this condition, inside this group, in this situation, for this client, and at this time?"

Faculty members were expressly asked to review the basic components that make up the core of their disciplines as part of the development of the IEP, in highlighting principles that could facilitate life-long learning. This necessitated admitting that a single degree program's time restrictions prevented thorough coverage of a discipline. The gap between the intent of the curriculum and learning outcome is generally too wide and it helps them to develop knowledge, insights, and problem solving skills. Furthermore, the recently defined appropriate parts of engineering were interwoven within core instruction, most typically within project-based elements, as recently specified by accrediting organizations. These socio-cultural and management themes were typically presented in distinct impartial components, which were repeatedly ignored by both faculty and learners students. Including such themes in the curriculum is not only more proficient, but it also helps students comprehend how their larger studies relate to their future careers as engineers.

7.15 EMBEDDED SKILLS

Professional groups representing engineers from various disciplines have incorporated transferable and management skills to their accreditation standards, in addition to technical skills, as seen above (e.g., IET, IMechE, REFS). Communication, leadership, teamwork, and resilience, as well as a stronger emphasis on sustainable and ethical engineering practice, have all been recognized by industrial and professional policy organizations as part of the "employability" package. Team projects are also excellent places to hone communication abilities, develop important interpersonal skills, and practice resilience and collaboration.

The entire joint effort strategy is aimed to facilitate students' individual expansion of collaboration processes by giving them a set of responsibilities to do that, when completed, support the functioning of productive teams [12]. Our team structure is, in many ways, a collection of review and support measures that our students will be able to put into practice at various stages all through their degree program. Over half of our first-year students said they learned the most from teamwork in their first team-based PBL activity.

7.16 PROMOTION OF INTRINSIC MOTIVATION
AND SELF-EFFICACY

Conceivably the IEP's primary idea is to support and inspire graduates; learners are innately motivated and do not require a set of direct methods from their tutors. The gold standard of student involvement is intrinsic motivation. Students who are internally driven are more liable to forfeit attention, like their effort, and adopt a position that promotes their self-efficiency and growth.

7.17   CONCLUSION

Higher education institutions must now follow a well-planned teaching/learning system that includes global-standard curriculum, course, instruction, and exam designs. In this environment, higher education teaching and research organizations must restructure their courses to meet OBE's needs.

REFERENCES

1. Rao, O.R.S. (2013). Outcome based engineering education – Need of the hour. Journal of Engineering Education.
2. Barab, S.A., & Duffy, T. (2000). From practice fields to communities of practice. In D. Jonassen & S.M. Land (Eds.), *Theoretical foundations of learning environments* (pp. 25–56). Mahwah, NJ: Lawrence Erlbaum Associates.
3. Barrows, H.S. (1992). *The tutorial process.* Springfield, IL: Southern Illinois University Medical School.
4. Brown, J.S., Collins, A., & Duguid, P. (1989). Situated cognition and the culture of learning. *Educational Researcher,* 18, 32–42. doi:10.3102/0013189X018001032
5. Buchanan, R. (1992). Wicked problems in design thinking. *Design Issues,* 8(2), 5–21. CBI. (2009). *Future fit: Preparing graduates for the world of work.* London: Author.
6. Allen, B. (2001). On the possibilities for optimal collaborative engineering design under uncertainty. In Proceedings of Optimization in Industry II-1999, New York: ASME Press.
7. Addis, W. (1990). *Structural engineering: The nature of theory and design.* Chichester, UK: Ellis Horwood.
8. Vesilind, P.A., & Gunn, A.S. (1998). *Engineering, ethics, and the environment.* New York: Cambridge University Press.
9. Ambrose, S.A., Lovett, M., Bridges, M.W., DiPietro, M., & Norman, M.K. (2010). *How learning works: Seven research-based principles for smart teaching.* San Francisco, CA: Jossey-Bass.
10. Harris, E.C, Pritchard, M.S., & Rabins, M.J. (1995). *Engineering ethics: Concepts and cases.* Belmont, CA: Wadsworth.
11. Herrington J., & Kervin, L. (2007). Authentic learning supported by technology: Ten suggestions and cases of integration in classrooms. University of Wollongong, Australia.
12. Atman, C.J., & Nair, I. (1996). Engineering in context: An empirical study of freshmen students' conceptual frameworks. *Journal of Engineering Education,* 85(4), 317–326.

Section IV

Applications

8 Information Transfer in Education

Case Study Based on Teaching Lowcost Control Unit Programming

Alena Hašková and Peter Kuna
Constantine the Philosopher University
Nitra-Chrenová, Slovakia

Miloš Palaj
Secondary Vocational School of Polytechnics
Zlaté Moravce, Slovakia

CONTENTS

8.1 INTRODUCTION

The rapid development of different kinds of digital technologies, their applications and innovations creates a basis for the formation of innovative digital economy (Maksaev, Vasbieva, Sherbakova, Mirzoeva, & Králik 2020), which is also incorporated into education (Billinghurst, Clark, & Lee, 2015; Maturkanič, Tomanová Čergeťová, Kondrla, Kurilenko, & Martin, 2021; Stojšić et al., 2020). Teachers are forced to seek innovative solutions to replace traditional teaching materials, methods and tools to fit into the world of digital technology (Galimova, Konysheva, Kalugina, & Sizova, 2019; Saienko, Kalugina, Baklashova, & Rodriguez, 2019). The ubiquitous digital environment and the amount of interaction with it make today's students think and process information in a fundamentally different way than their predecessors. The students' approach to learning is starting to be characterized as a trial-and-error process rather than systematic assessment. Therefore, according to Prensky (2001), in order to maximize the attention and efforts of students, information should be presented in a way that is best suited for this approach – based on the use of different possibilities given by the new digital technologies. Yet, the new digital economy and the platform of the fourth industrial revolution connected to it – a concept presented by the Government of the Federal Republic of Germany at the Industrial Fair at Hannover in 2013 (BMWi, 2016) – require not only the innovation of the educational process by the implementation of digital technologies but also the content of education to be innovated, too. Therefore, it can be concluded that the connection of the digital economy with Industry 4.0 requires a system for working in cooperation, not just "digitizing" processes. It requires an integrated approach to education and vocational training with new goals, structure and content to achieve a high level of professional competencies required by the labour market of the twenty-first century (Kobylarek, 2019, 2018; Maksaev et al., 2020). One of the fastest growing fields in the labour market within the environment of Industry 4.0 is in the field of programming, which needs adequately trained and skilled programmers.

Following upon the aforementioned context we have focused our attention on issues of vocational education and the training of programmers. We were interested in the question of how the transfer of professional information is processed in their case and whether there are some specific aspects of this process compared with, for example, the transfer of information carried out within general education. Hereinafter we present results of the case study, whose aim was to analyse the specifics of learning and acquiring information carried out in the framework of teaching secondary school students to programme lowcost control units.

8.2 BACKGROUND OF THE RESEARCH

Experts dealing with the issues of learning and information transfer emphasize that even getting familiar with the subject matter undoubtedly takes time, so it is very important how learners use their time while learning. In this context, it has been generally accepted that learning with understanding is more likely to support information transfer than simply memorizing information from a text or a lecture. However, a lot of learning activities are based on memorizing rather than learning for understanding. This means that familiarization with facts and details is in the centre of attention rather than the causes and consequences of events. Knowledge taught in a variety of contexts is more likely to support flexible transfer than knowledge taught in a single context (Bransford, Brown, & Cockin, 2000). Multiple contexts support the development of a more flexible understanding of when, where, why, and how to use the acquired knowledge to solve new problems (known as conditions of applicability).

According to Singley and Anderson (1989), if information transfer is to be successful, it has to be based on previous experience and requires initial learning, as well. Besides that, different kinds of knowledge representations can support information transfer and enhance its effectiveness. Teachers have a critical role in assisting their learners in building on understanding, correcting misconceptions, and observing and engaging with the learners during the process of learning.

As to information transfer, Bransford, Brown and Cockin (2000) draw attention to the fact that the efficiency of information transfer can be increased by means of using different transfer categories of the same information. This means that different ways of repeated transfer of the same information cater for complex cognitive information processes, which the brain activates at a higher level. In this way neuroscience (Petlák & Trníková, 2010) confirms one of the basic principles presented in Komenský's (Comenius') *Didactica Magna – Great Didactic* (Komenský, 1954, 1948). It is the rule that information should not be mediated to learners only by words but also by pictures, physical objects, video and audio records, demonstrations and by doing activities. Moreover, Komenský pursued teaching directed by experience, because he believed that children acquire wisdom by sensory observations and experiences (Jůvová & Bakker, 2015; Lukaš & Munjiza, 2014). Here there is consensus with the findings of neuroscience, which confirms the important role that experience plays in building the structure of the mind by modifying the structures of the brain, regardless of what is remembered. During the transfer the brain is activated by means of different categories of information, such as sounds, words uttered out loud, written texts and pictures, as well as by means of new questions arising and answers forming related to the issues studied. Experience is important for the development of brain structures and what is registered in the brain is connected with its own mental activities (Schacter, Gilbert, Wegner, & Nock, 2015, Schacter et al., 2012; Wagner, Maril, & Schacter, 2000). Educators face the challenge of making knowledge and information transfer as efficient as possible. As research comparisons of people's memories for words with their memories for pictures of the same objects show, there is a superiority effect for pictures (Brown, Roediger, & McDaniel, 2014; Roediger, 2008; Roediger, Dudai, & Fitzpatrick, 2007). Simultaneously it has been proven that even knowledge built on abstract representation does not remain isolated as the mind tries to incorporate it into some larger related phenomena (Biederman & Shiffrar, 1987; Holyoak, 1984; Novick & Holyoak, 1991).

The effectiveness of information transfer is the subject of research in both the neurosciences and general didactics. Subject didactics only seldom deal with this issue even though transfer effectiveness is closely related to the mediated content, which, in our case is the type of the subject matter. In the case of programming, informal interviews with students indicated that education in the field of microcontroller programming (or programming in general) is probably specific in that the information provided through video tutorials is not as effective and preferred from the students' point of view as in other fields of study (e.g., humanities). Thus, an interesting paradox arises in education in the field of programming lowcost control units. Obst (Kalhous, Obst et al., 2002) claims that the individual learner can learn new information more quickly when it is presented through video than when they have information conveyed in the form of textual and pictorial instructions. In our case (i.e. in the field of programming lowcost control units), however, it seems that textual and, especially, visual instructions stick more than information conveyed by video.

8.3   PURPOSE AND RESEARCH QUESTIONS

The aim of our case study was to set up the optimal way of education in the field of Arduino microcontroller programming. Optimality was monitored in relation to the various ways of implementing the transfer of technical information (or curriculum) to the students (Kuna, Palaj, & Hašková, 2019). To this end, it was necessary to find an answer to the question of what form of providing information to students in teaching programming is the most appropriate in terms of its effectiveness.

The search for the answer to this research question was based on the design of alternative ways of teaching the identical programming content of lowcost control units. The alternative teaching methods were based on different ways of transferring information to the students: via text, pictures and video recordings.

Programming lowcost control units is taught at secondary vocational schools with a technical major, mostly within the practical lessons of Arduino and single-chip microcontrollers thematic units, the teacher having 33 hours per year in total for this subject (one hour per week). In accordance with the yearly total time allowance of practical lessons the curriculum has been grouped into 16 topics, each in the time allowance of a two-hour lesson:

1. Technical details of Arduino
2. Programming simple applications
3. Programming simple applications
4. Programming analogous inputs
5. Programming more complex applications
6. Programming more complex applications: connecting a DC (direct current) motor and a DC motor control
7. Control of a stepper motor
8. Servo motor control
9. Arduino, relay switching elements and transistor
10. Connecting the electronic components to Arduino (NE555)
11. Connecting electronics components to Arduino (shift register)
12. Arduino and wireless devices
13. Arduino and sensors I
14. Arduino and sensors II
15. Arduino and sensors III
16. Arduino and sensors IV

8.4   METHODOLOGY OF THE RESEARCH AND PARTIAL RESEARCH QUESTIONS

The effectiveness of the individual teaching methods was subsequently verified in practice using a pedagogical experiment, which was based on the comparison of the course and the results of three alternative teaching methods. A total of 24 students were involved in the experiment, who were divided into three eight-member groups. All three groups were de facto experimental. In the first group, the teaching process was based on written instructions. The second group of students had access to the instructions of the exercises, where the content was explained using text and pictorial instructions. Let us note that it was the same content of exercises as in the first case, only the form – the way of presenting the information – was altered. The third group of students had to perform the same tasks as the two previous groups, but in the case of this group it was done via video instructions.

The content of the exercises was presented to the students on a specially created educational website (Figure 8.1). The exercises were made available to the students as practical step-by-step lessons. Each group had access only to their respective form of mediated curriculum secured by a password. At the end of each exercise the students (in each group) were given a test.

To be able to find an answer to our research question based on the presented research concept, it was necessary to answer the following partial research questions:

1. How do the students perceive the difficulty of the units?
2. How do the students perceive the content of the exercises?
3. How do the students perceive the sequence of the units?
4. How do the students perceive teaching programming lowcost units via written instructions?

FIGURE 8.1   Menu of the exercises (for the second group) on the created educational website.

5. How do the students perceive teaching programming lowcost units via written and pictorial instructions?
6. How do the students perceive teaching programming lowcost units via video tutorials?

As a tool of collecting research data, a semi-structured interview was used on the basis of which it was be possible to evaluate why students have a positive or negative standpoint regarding a certain way of mediating information and write down their attitudes in connection with the nature of the acquired knowledge (Hendl, 2005; Silverman, 2013). The topics we focused on during the interviews were related to finding answers to the above-mentioned partial research questions. If the students in the groups agreed on their answers to the researched questions in at least five of eight cases, we considered their answers relevant and guiding. We considered a lower number of identical answers to be insufficient.

8.5 RESULTS

In the following, we present the results of the implemented pedagogical experiment summarized according to the respective partial research questions.

8.5.1 How Do the Students Perceive the Difficulty of the Units

In the first two exercises the students from all the three groups considered the difficulty of the exercises to be less demanding regardless of in what form the information was provided. The third and fourth exercises were considered to be relatively simple too. Difficulty increased when it came to connecting the microcontroller terminals, where it was necessary to use more wires. The students considered the fifth exercise to be slightly demanding, as it was necessary to add libraries to the circuit and the program, which enable communication with the LCD display. The sixth to eighth exercises, focusing on motor programming, were assessed by the students as moderately demanding comparable to the fifth exercise. The students had mild problems understanding the functionality and programme part by which the motors are controlled. The ninth exercise, focusing on switching elements, was considered to be less demanding than the units on motor control. According to the students, the difficulty and programming part of this exercise was similar to that of the third and fourth exercises. In this exercise the students did not need to use additional libraries to control the switching elements. It was enough to use only a few lines of the program, which the students understood rather well. The tenth and eleventh exercises were considered to be more demanding especially from the aspect of programming and the practical part. The students had to use additional libraries to make it work, which accounted for the higher complexity of exercises. In order to do that, they also needed to connect a larger number of wires to the microcontroller and, last but not least, write a relatively demanding program. In addition, they had to use knowledge gained in another subject to be able to do these circuits correctly. This fact also accounted for the more demanding nature of the exercise. The students considered the twelfth exercise to be very demanding both from the aspect of programming and the practical part. The students focused on wifi and bluetooth modules that enable remote control of peripheral devices. They considered the overall implementation of the connection to be the most demanding as it was necessary to correctly define the libraries, understand the program and write it correctly. In addition, it was necessary to use interdisciplinary relationships (commands, so-called tags for creating web pages). The thirteenth to fifteenth exercises focused on sensing elements. The students considered these exercises to be slightly demanding in terms of content, especially in the programming part. In terms of complexity, they considered the exercises in the practical part to be easier because a low number of drivers (five on average) was sufficient to ensure proper functionality. The main circuit consisted primarily of the programme part, which was supposed to primarily ensure the proper working of the sensors. It was also important to use the right type of libraries. If the students used the wrong type of library, the connection did not work properly or not at all. Overall, the students considered the thirteenth to fifteenth exercises to be slightly demanding but certainly easier than the tenth to twelfth exercises. The last exercise, which focused on the NFC sensor and motion sensor, was considered rather challenging by the students. Especially the programming part was hard for them, where they had to use a larger number of libraries and define more variables than in the thirteenth to fifteenth exercises.

8.5.2 How Do the Students Perceive the Content of the Exercises

The students considered the first and second exercise to be a bit boring for them; nevertheless, it met their expectations. They appreciated the detailed description of the microcontroller due to which they understood the principle of how this platform operates. They considered the third and fourth exercises to be sufficiently extensive in terms of content due to the length of time allocated to the exercises. They only needed to look at the lesson once to understand the main idea of the exercises. They appreciated the part from the content page, where the functionality and behaviour of analogue outlets was pointed out using a few examples. The fifth exercise, and within it the description of the programme, was considered sufficient in terms of content and very well explained as to the functions of the individual parts. According to the students, the programming of the LCD display was sufficiently explained, but in some parts, there was a lot of description and therefore they felt they did not have enough time to complete the exercise. The sixth to eighth exercises were considered to be too extensive and inadequate in terms of content by the learners. According to them, it would be enough to use simpler vocabulary and use only one example. However, according to the school curriculum, it is not possible to modify the content and still meet the standards. By reducing the complexity, we would not sufficiently cover all the necessary aspects of motor control via a microcontroller. Content-wise, the ninth exercise was perceived as comprehensive and sufficient, with no need to add additional tasks related to relay programming and transistor switching elements. The students considered the tenth and eleventh exercises to be extensive and demanding. Nevertheless, it is not possible to reduce the content here either because, as in the case of the sixth to eighth exercises, by reducing the complexity we would not cover all the aspects of programming NE555 add-on circuits and shift registers required by the training program. The students considered the twelfth exercise to be sufficient but demanding in terms of content. They accepted the range of content and realized that if there were less content they would not understand all the aspects of programming wifi and bluetooth modules. In the thirteenth to fifteenth exercises, according to the students, adequate attention was paid to each sensor, so it was not necessary to supplement the content with further demonstrations and explanations when doing the exercises. The sixteenth exercise was also considered sufficient but demanding in terms of content. As in exercise 12, the students accepted the scope of the content and realized that, to a lesser extent, they would not understand all the context of programming.

8.5.3 How Do the Students Perceive the Sequence of the Units

The students positively evaluated the sequence of the third and fourth exercises, where the units deal with analogous parameters, because in order to understand the analogous inputs of the microcontroller it is necessary to learn how to use the digital inlets correctly and understand how to program them. The fifth exercise, dealing with the connection and programming of the LCD display, the students evaluated positively, especially the part where they had to use the knowledge of connecting analogous parameters. The sixth to eighth exercises, dealing with the connection and programming of small motors, were evaluated positively too, in particular for the sequence of the units, because in order to understand the control of the direct current it is necessary to learn how to use the digital outlets. However, the eighth unit, dealing with servomotor programming, they evaluated less positively, as the students understood the programming of the servomotor better than that of the stepper motor. For this reason, they proposed to include the unit devoted to the stepper motor after the unit focusing on servomotor programming. In case of the ninth exercise, focused on the switching elements of relays and transistors, the sequence of the units was evaluated rather negatively. The students proposed to include this topic before programming the DC motor. Sequences of the tenth and eleventh exercises were evaluated positively as, in order to understand the operation of the NE555 integrated circuit and shift registers, it is necessary to learn how to use the libraries that the students learned to use in the third exercise. The twelfth exercise, focusing on programming wifi and bluetooth modules, was challenging for the students. The sequence of the unit was assessed negatively. According to the learners, it would be appropriate to include this topic as the last one, divide it into more modules and deal with each of them separately. Related to this was the short time allowance, which the students proposed to increase. The sequence of the thirteenth to fifteenth exercises, focused on sensing elements, was evaluated less positively, as the students considered them simpler and suggested moving them before the unit with NE555 circuits and shift register sensors. The students evaluated the sequence of the sixteenth unit focusing on NFC and motion rather positively, but as they considered it only slightly demanding, they suggested moving it after the unit with NE555 circuits and a shift register. The students also proposed dividing this unit into two separate ones.

8.5.4 How Do the Students Perceive Teaching Programming Lowcost Units via Written Instructions

The students educated exclusively in written form during the first two exercises considered the mediated content to be sufficiently clear. The text form suited them, but for a better understanding of the final scheme they would appreciate at least one picture of the relevant circuit. The third and fourth exercises were evaluated similarly to the first two ones. According to the learners the curriculum was generally presented in an understandable way. However, students had reservations about the part of the curriculum where they had to connect and program analogous inputs. They would appreciate a more detailed description of the process of connecting the individual conductors or would love to see one pictorial manual showing the final scheme. In the case of the fifth exercise, due to the text format, the students considered the content to be slightly demanding. Especially when understanding the principles of functioning and connecting the LCD display to the microcontroller the written form of providing information was not sufficiently clear to them. The problem occurred mainly when connecting the display, because for its proper functioning it is necessary to connect a larger number of conductors and use libraries that allow for the correct functioning of the LCD display. The students considered the sixth to eighth exercises to be slightly demanding. The reason given was mainly the complexity of the technical connection, where they could not sufficiently imagine the scheme. The students had reservations mainly about the part of the curriculum where they had to connect the inputs themselves. They would appreciate a more illustrative way of connecting the individual parts (e.g., in the form of visual or video instructions). The students understood the ninth exercise but would consider the provision of non-textual information. They said the theoretical information as well as the explanation of the principles of operation were more or less clear but they had reservations about the part of the curriculum where they had to physically connect the transistor and the relay. Here they would appreciate another form of providing information. In the case of the tenth and eleventh exercises, the students only partially understood the curriculum (presented in text format). They missed the pictorial or video instructions. They claimed that the curriculum was mediated rather incomprehensibly. Although the written information was sufficient, the students had no idea how to properly connect the individual pins as it was a relatively complex integrated circuit. In this context they would appreciate a more detailed description of connecting the individual wires to the integrated circuit. As far as the twelfth exercise is concerned, the students understood it very little or not at all. The reason for it was mainly the absence of pictorial and video instructions as a form of transmitting information and also the large content of the information conveyed. Although the textual information was sufficient and the students understood the theoretical information, it did not give them any idea how to connect the individual components. The students only marginally understood the practical part of the thirteenth to fifteenth exercises. They would consider the presence of pictorial or video instructions as possible forms of providing information to be much more illustrative. According to them, the curriculum was mediated rather incomprehensibly. They had reservations mainly about the part of the curriculum where they had to connect and program each sensor inlet.

They would appreciate a more detailed description of the procedure of connecting the individual wires. On the other hand, they considered the theoretical knowledge in this area to be comprehensive and sufficient as they understood the theoretical context and functionality of these sensors. They considered the sixteenth exercise to be more demanding and they did not understand the practical part very much. They would consider pictorial or video instructions as forms of providing information much more illustrative. However, from a theoretical point of view, they had no problems understanding the functionality of the sensors. They considered the theoretical knowledge in this unit to be sufficiently comprehensive, enabling them to understand the functionality of these sensors.

8.5.5 How Do the Students Perceive Teaching Programming Lowcost Units via Written and Pictorial Instructions

The students who received instructions in written and pictorial form did the first two exercises fully independently. The formal side of the exercises was evaluated positively by all of them and they also highlighted the visual aspect of it, which enabled them to have a look at the individual parts of the device. They especially appreciated the illustrative demonstration where they could see a detailed description of the microcontroller. They considered the video tutorial unnecessary in this case. In the third and fourth exercises the written text was considered comprehensible, sometimes too detailed. This impression was caused mainly by the fact that they also had a visual guide at their disposal, from which they were able to pick up the principle of programming analogous inputs and programming a timer. The interviews with the students of the second group showed that they skipped the written part and followed the pictorial instructions only. As a result, they did not master the theoretical foundations and the basic principles of the operation of the circuit. In the fifth exercise the students positively evaluated the formal side of the exercise. They assessed the pictorial instructions as clear and illustrative. Five students rated the time allowance as sufficient, were able to work independently and completed all the assigned tasks. They considered the programming part to be understandable and understood it the first time. Two students needed slight help from the teacher and one student needed more help. On the other hand, the students had a slightly weaker theoretical knowledge as they skipped the written description in this exercise, too, and went straight to connecting the circuit. The students managed the sixth to eighth exercises similarly to the previous ones. They considered the written text to be comprehensible, sometimes too detailed. This impression was mainly caused by the fact that they also had a pictorial manual available, from which they were able to understand the connection and operation of the DC motor more quickly than the students in the first group. However, they could not imagine the activity of motors and therefore would appreciate a video tutorial with a demonstration of their activity. The ninth exercise was mastered by the second group similarly to the first one, mainly due to the fact, that the students also had a pictorial manual from which they were able to more quickly understand the principle of connecting and programming the relay or transistor. The students positively evaluated the clearly presented visual information which was mostly used when doing the connections. In this exercise the students understood the program at first sights. Two students needed some help from the teacher because it was necessary to define a different microcontroller terminal. In this case the students skipped the written instructions again and started connecting straight. This resulted in a poorer understanding of the functionality of the circuit. The tenth and eleventh exercises were considered as moderately demanding. The students considered the written text to be understandable, but sometimes unnecessary in support of the pictorial instructions. This impression was mainly caused by the fact that they also had a pictorial manual at hand, from which they felt that they could understand the principle of integrated circuit programming and did not need a written manual. The students positively evaluated the clearly presented visual information, which was mostly used during connecting. In this exercise the students understood the program at first glance. Two students needed some help from the teacher, as it was necessary to define other elements needed for the connection (correct connection of the integrated circuit power supply). In this case, however, eight students did not pay enough attention to the textual form of the information provided but focused mostly on the pictorial form. This resulted in the students not understanding the functionality of the circuits and insufficient theoretical knowledge. The twelfth exercise was managed very similarly to the previous exercises. The students considered the written text to be clear, sometimes unnecessary, in support of the pictorial instructions. They justified it by saying that they also had a visual guide at their disposal, from which they were able to understand the principle of programming this circuit (faster than the students in the first group). In the end, however, they were wrong because they performed worse in the test than the first group. Therefore, we consider textual information to be an important element of education, especially when it comes to the level of the students' knowledge. On the downside, the students stated that the presence of the pictorial instructions greatly reduced the importance of the textual description. This resulted in a poorer understanding of the use and technical behaviour of the circuit. Seven students positively evaluated the clearly presented visual information, which they mostly used during connecting. In this case, however, the students did not pay enough attention to the textual form of the information and paid attention mostly to the pictorial form. The students found the thirteenth to fifteenth exercises, focusing on the sensing elements, to be moderately demanding. They considered the written text to be intelligible in support of the pictorial instructions, but in six cases the students skipped the textual description, which was supposed to help acquire theoretical knowledge and moved on to the practical schemes and connecting. The results of the test in this group subsequently pointed to a lower level of acquired knowledge. The sixteenth exercise was managed again very similarly to the previous two exercises. The students skipped the written part, went straight to the connection of the schemes and subsequently achieved worse results in the test.

8.5.6 How Do the Students Perceive Teaching Programming Lowcost Units via Video Tutorials

The students educated exclusively in the form of video instructions evaluated the form of education of the first two exercises very positively. The content was clear for them and they understood the curriculum, which was also proved by the test results.

When asked about their opinion about the form of education via video instructions, six students expressed a positive attitude, appreciating the clarity of information provided. The other two students considered the video guide to be confusing because they had to watch it repeatedly and missed some of the parts needed for the connections. All eight students appreciated the third and fourth exercises mainly from the illustrative and content point of view as they saw the specific procedure and method of connecting. The way of teaching was evaluated very positively by seven students and less positively by one. What the students did not like was the fact that they had to replay the video and pause it at certain parts so that they could do what they saw in the video. For this reason, they wasted time and two of them did not manage to do all the necessary tasks. In the fifth exercise, which focused on the connection of two-line LCD displays, they evaluated the formal side of the exercise as interesting and understandable. Five students complained about their inability to copy parts of the program as it was a relatively long one in terms of the LCD display and the libraries that were needed to make it work. As a result, seven students had to repeatedly pause the video and play it back multiple times and three students needed help from the teacher with doing the connections. They would prefer a written tutorial instructing them about the wiring process. They considered the visually presented procedure to be unnecessary because the video gave them a good enough overview of the connections. The students appreciated the sixth to eighth exercises presented by the video mainly from the illustrative and content point of view. The form of mediating the curriculum was evaluated very positively by five students, less positively by two and insufficiently by one. The disadvantage for the students in this exercise was the need to replay the video and pause it at certain parts so that they could copy the processes. This way they wasted time and three students did not manage to finish all the necessary tasks. In the ninth exercise the students appreciated the presentation of the curriculum mainly from the illustrative and content point of view, as they saw the specific process and wiring in the form of video. The form of mediating the curriculum was very positively viewed by seven students and less positively by one of them. The disadvantage for the students was again the need to replay the video and pause at certain parts so that they could copy the steps from the video, which resulted in a higher error rate during wiring. Similarly, in next two exercises, the students appreciated the curriculum processed in the form of video mainly from the illustrative and content point of view (seven very positive and one less positive assessment). The students viewed the explanations and the provision of the theoretical information about the principles of NE55 integrated circuits via video instructions rather negatively, as a matter of fact, they assessed this way of learning as essentially negative. The twelfth exercise was again appreciated by the students mainly from the illustrative and content point of view, because they could see the procedure, workings and wiring in the video. Despite the positive evaluations the students described the video form of giving information as boring. Due to repeated pausing and replaying three students did not have time to complete all the necessary assignments. They would therefore prefer the thematic unit to be divided into two smaller parts, where they would deal with each device separately. Explaining the principles of integrated circuits via video instruction was perceived by the students as an inappropriate way of learning. Although they understood the laws of how the circuit operates, they only partially understood the principles of its programming. The theoretical knowledge gained regarding the operation of these modules was perceived as insufficient, as the students skipped the part of the video manual where this type of information was provided and went directly to connect the circuit. Learning with a video during the next three exercises was evaluated very positively by all the students of the third group except for one. Seven of the students managed the wiring on time and one did not manage to finish up all the connections. The students considered the acquisition of theoretical knowledge through video to be less useful, since they deducted the principles from the functions of the sensors. The test carried out at the end of the exercise, however, pointed to the students' shortcomings in understanding the principles of how the sensors work. The programming part was understandable for them as the main part of the program consisted of additional libraries which they were already using. The sixteenth exercise was evaluated more positively: the way the curriculum was mediated was very positive for four of the students and less positive for the other four. Three students did not manage to finish up all the necessary tasks during the exercise, but five of them managed to get the connections done independently and on time.

8.6   CONCLUSIONS

The presented results of the implemented research point to the fact that students lack clarity if education is mediated exclusively in written form. Overall, this method of education (applied in the case of the first group) was considered as lifeless or boring by the students. The students could not imagine the connections and functions of the circuits well enough during the exercises. At some point during the experiment, we tended to make premature conclusions that teaching programming lowcost control units via written instruction is insufficient, perhaps even inappropriate, and students should not be educated this way. Later, during the analysis of the recorded semi-structured interviews and the results of the tests, we found out that, although the students had problems with the practical connections of the circuits, they acquired the necessary theoretical knowledge needed to understand functionality and find relationships between the various types of circuits (within one lesson as well as within the context of the previously acquired knowledge). The theoretical knowledge of the students of the first group clearly exceeded that of the students of the second and third group.

The students of the second group both had written and pictorial instructions. The findings of the research point to the fact that these students focused mainly on the pictorial instructions and skipped the text-based learning materials. As a result, they did not acquire the necessary knowledge and they were not aware of the various important contexts such as the theoretical knowledge of the circuit, the principles of its workings or the way of programming. As a result, the students' learning outcomes reached the required level only in terms of acquiring the mechanical manual skills ("the automatic" connection of the circuit parts based on "reading" the presented scheme without conscious knowledge of the principles and awareness of the operation of the individual components of the scheme involved). At the knowledge tests these students showed weaker results (a lower number of points were obtained) than the students in the first group. It seemed as if the exercises had been aimed at drilling and developing the skills of connecting the relevant circuits, which was not the case. Based on the presented results it may be argued that visual materials (diagrams) are clearly a suitable form of providing information when it comes to doing the practical connections of circuits but are insufficient for acquiring theoretical knowledge.

The students of the third group were educated via video instructions, which allowed them to see and understand the workings of the individual circuits. From the illustrative point of view, the students rated the video instructions positively. On the other hand, they did not manage to mentally process the presented facts and contexts and had to pause the recordings or replay certain parts of the video being watched. As a result, this way of learning the curriculum was lengthy and boring for students. They solved the problem of length and boredom by not spending enough time and not paying sufficient attention to the principles of circuit functions, skipping the relevant parts of the video instructions and focusing only on the connection of circuits. This was reflected in their poorer understanding of circuit functionality and lower success in the tests (compared to the students in the first group). Nevertheless, the video tutorial proved to be an excellent and suitable way of teaching small motor programming. At first glance, it might have seemed that the video tutorial would be the most appropriate way to teach. However, it turned out that when programming lowcost control units it is appropriate to use it as an illustrative demonstration activity at the end of the lesson rather than as the central means of education. Only in the case of the units focusing on small motor programming can it be recommended as the main way of conveying information.

The presented results of our case study show that information provided via video tutorials is not the most suitable way of teaching programming lowcost control units. This goes against the claim that video is the strongest stimulant for learning and remembering information (Kalhous, Obst et al., 2002). Nevertheless, we are aware of the fact that the above findings cannot be applied to all subjects but only to that one which focuses on programming lowcost control units. The results of the presented study show that it is recommended to use a combination of text and pictures to teach this field of study. What is more, they serve as a basis for deciding on the correct mix or blend of these two ways of teaching.

REFERENCES

Biederman, I., & Shiffrar, M. M. (1987). Sexing day-old chicks: A case study and expert systems analysis of a difficult perceptual-learning task. *Journal of Experimental Psychology: Learning, Memory, and Cognition*, 13(4), 640–645. doi:10.1037/0278-7393.13.4.640.

Billinghurst, M., Clark, A., & Lee, G. (2015). A survey of augmented reality. *Foundations and Trends in Human–Computer Interaction*, 8(2-3), 173–272. doi:10.1561/1100000049.

BMWi, Federal Ministry for Economic Affairs and Energy. (2016). *Digitization of Industrie – Platform Industrie 4.0*. Available at https://www.plattform-i40.de/PI40/Redaktion/EN/Downloads/Publikation/digitization-of-industrie-plattform-i40.pdf?__blob=publicationFile&v=4.

Bransford, J. D., Brown, A. L., & Cockin, R. R. (Eds.). (2000). National Research Council: *How People Learn: Brain, Mind, Experience, and School*. Expanded Edition. Washington, DC: The National Academies Press. Doi:10.17226/9853.

Brown, P. C., Roediger, H. L, & McDaniel, M. A. (2014). *Make it stick: The science of successful learning.* Cambridge, MA: Harvard University Press. ISBN 0674729013.

Galimova, E. G., Konysheva, A. V., Kalugina, O. A., & Sizova, Z. M. (2019). Digital educational footprint as a way to evaluate the results of students' learning and cognitive activity in the process of teaching mathematics. *Eurasia Journal of Mathematics, Science and Technology Education* 15(8), article no. em1732, 10 p. doi:10.29333/ejmste/108435.

Hendl, J. (2005). Kvalitativní výzkum: základní metody a aplikace qualitative research: fundamental methods and applications. Praha: Portál.

Holyoak, K. J. (1984). Analogical thinking and human intelligence. In Sternberg, R. J. (Ed.), *Advances in the Psychology of Human Intelligence,* (Vol. 2, 199–230). Hillsdale, NJ: Erlbaum.

Jůvová, A., & Bakker, F. (2015). Didactic principles by Comenius and 21st century skills. *E-Pegagogium*, II(2), 7–20. doi: 10.5507/epd.2015.013.

Kalhous, Z., Obst, O. et al. (2002). Školní didaktika. Praha: Portál. ISBN 80-7178-253-X.

Kobylarek, A. (2018). Science as a bridge. Science in action preface. *Journal of Education Culture and Society*, 9(2), 5–8. doi: 10.15503/jecs20182.5.8.

Kobylarek, A. (2019). Social responsibility of science introduction. *Journal of Education Culture and Society*, 10(2), 5–11. doi: 10.15503/jecs20192.5.11.

Komenský, J. A. (1948). *Didaktika velká.* Brno: Komenium.

Komenský, J. A. (1954). *Velika didaktika.* Beograd: prijevod u izdanju Saveza pedagoških društava Jugoslavije.

Kuna, P., Palaj, M., & A. Hašková, A. (2019). Alternative ways of teaching programming of lowcost control units (ARDUINO). In EDULEARN19 Proceedings, 1627–1632. Valencia: IATED Academy. ISBN 978-84-09-12031-4, ISSN 2340-1117.

Lukaš, M., & Munjiza, E. (2014). Education system of John Amos Comenius and its implications in modern didactics. *Život I škola*, 31(1), 60(31), 32–44. Available at https://eric.ed.gov/?id=ED556095.

Maksaev, A. A., Vasbieva, D. G., Sherbakova, O. Yu., Mirzoeva, F. R., & Králik, R. (2020). Education at a cooperative university in the digital economy. In A. V. Bogoviz, A. E. Suglobov, A. N. Maloletko, O. V. Kaurova, & S. V. Lobova, (Eds) *Studies in Systems, Decision and Control* (vol. 316: *Frontier Information Technology and Systems Research in Cooperative Economics*, 33–42). Cham: Springer Nature Switzerland AG. doi: 10.1007/978-3-030-57831-2_4.

Maturkanič, P., Tomanová Čergeťová, I., Kondrla, P., Kurilenko, V., & Martin, J. G. (2021). Homo culturalis versus cultura animi. *Journal of Education Culture and Society*, 12(2), 51–58. https://doi.org/10.15503/jecs2021.2.51.58.

Novick, L. R., & Holyoak, K. J. (1991). Mathematical problem solving by analogy. *Journal of Experimental Psychology: Learning, Memory, and Cognition*, 17(3), 398–415. doi: 10.1037/0278-7393.17.3.398.

Petlák, E., & Trníková, J. (2010). *Neurodidaktika a vyučovanie: Úvod do problematiky mozgovokompatibilného učenia.* München: GRIN Verlag. ISBN 3640686004.

Prensky, M. (2001). Digital natives, digital immigrants. *On the Horizon* 9(5), 1–6, MCB University Press. doi:10.1108/10748120110424816.

Roediger, H. L. (2008). Cognitive psychology of memory. In J. Byrne (Ed) *Learning and memory: A comprehensive reference* (Vol 2). Oxford: Elsevier. ISBN 978-0-12-370509-9.

Roediger, H. L., Dudai, Y., & Fitzpatrick, S. M. (Eds.). (2007). *Science of memory: Concepts.* Oxford: Oxford University Press. https://doi.org/10.1093/acprof:oso/9780195310443.001.0001.

Saienko, N., Kalugina, O. A., Baklashova, T. A., & Rodriguez, R. G. (2019). A stage-by-stage approach to utilizing news media in foreign language classes at higher educational institutions. *XLinguae* 12(1), 91–102. doi: 10.18355/XL.2019.12.01.07.

Schacter, D. L., Addis, D. R., Hassabis, D., Martin, V. C., Spreng, R. N., & Szpunar, K. K. (2012). The future of memory: Remembering, imagining, and the brain. *Neuron, 76,* 677–694. doi: 10.1016/j.neuron.2012.11.001.

Schacter, D. L., Gilbert, D. T., Wegner, D. M., & Nock, M. K. (2015). *Introducing psychology,* 3rd Ed. New York: Worth. ISBN-13: 978-1464107818. ISBN-10: 1464107815.

Silverman, D. (2013). *Doing qualitative research a practical handbook.* 4th ed. London: Sage Publications. ISBN 978-1-4462-6014-2.

Singley, K., & Anderson, J. R. (1989). *The transfer of cognitive skill.* Cambridge, MA: Harvard University Press. ISBN 9780674903401.

Stojšić, I., Ivkov-Džigurski, A., Maričić, O., Stanisavljević, J., Milanković Jovanov, J., & Višnić, T. (2020). Students' attitudes toward the application of mobile augmented reality in higher education. *Društvena istraživanja,* 29(4), 535–554. doi:10.5559/di.29.4.02.

Wagner, A. D., Maril, A., & Schacter, D. L. (2000). Interactions between forms of memory: When priming hinders new episodic learning. *Journal of Cognitive Neuroscience,* 12(1), 52–60. doi:10.1162/089892900564064.

9 Home Assignments and Assessments in Online Learning

B. Nagamani, N. Subadra, S. Lalitha,
and Bhartipudi Saketh Ram

Geethanjali College of Engineering and Technology
Hyderabad, Telangana, India

CONTENTS

DOI: 10.1201/9781003083160-13

9.1 INTRODUCTION

Outcome-based education (OBE) is not a new term or a new movement. It has been present in the field of learning and education for 30 to 40 years in some form or another. However, it never gained momentum as it has gained now, primarily because its early understanding and elucidation were through rigid and complex formats; and poor or lack of involvement of teachers might have been a peripheral reason for causing it to fade away for some time. Let us try to understand OBE in simpler terms. It is an educational model in which curriculum, pedagogy, and assessments are interlinked to each other to ensure that the students know why they are learning and what are they learning. To be precise, everything helps in achieving student learning outcomes. OBE is an educational model that promotes learning and individual and institutional accountability based on student learning. In nutshell, it promotes learning for learning's sake. OBE gives abundant opportunities to create learner-centered classrooms. Pedagogical practices such as assessments play a major role in achieving learning outcomes.

In OBE the teacher's creativity is not restricted. In fact, it widens the horizons of the teacher's innovative methodologies to engage more learners in meaningful discussions and classroom participation in explaining the concepts that assess students' learning. Students' progress in learning and fulfilling learning outcomes is measured by achieving learning outcomes. In everyday language, outcomes are stated expectations.

Outcomes provide opportunities for deep learning to understand the concept better. OBE provides adequate platforms and opportunities to cultivate and promote deep learning rather than surface learning. Deep learning enables a learner to know the concept better, dig deep, research, and question, and finally it provides opportunities for expanded learning. Deep learning expands one's knowledge. Considering the advantages deep learning has over superficial or surface learning, the majority would accept that deep learning is intended and preferable.

The process of acquiring, analyzing, documenting, and applying information regarding students' performance on educational tasks is known as assessment (Lambert and Lines, 2000). Assessment serves a variety of purposes. Firstly it provides feedback to both teachers and learners. This helps the teacher and the student to change their future teaching and learning strategy. It also provides feedback to the parents about their ward's performance, regularity and other aspects from time to time. Secondly, it provides information about the student's progress. Thirdly, it provides a means of selecting students for further courses and/or employment. Finally, is assists in judging the effectiveness of the educational system. For all learning and classroom purposes, having a clear understanding of these multiple and diverse aims and whether they may be blended successfully is very crucial.

Assessment occupies a fundamentally important place in education. As it is an important aspect in teaching learning process, it should not be one dimensional. Assessment is the process of gathering and discussing information from a multitude of sources in order to gain a comprehensive understanding of what students know, understand, and can do with that knowledge as a result of their educational experiences. It concludes when the evaluation results in improving future learning (Huba & Freed, 2000, p. 7, emphasis added). Second, Palomba and Banta (1999) defined outcome-based assessment as "the systematic collection, review, and use of information about educational programs undertaken for the purpose of improving student learning and development" (p. 4).

In order to get a reliable, valuable, and complete understanding of the students' learning curve, it is essential that the type of assessment should be varied in order to meet the needs of individual learners. Assessment can be formative or summative. Summative assessment generally occurs at the end of the semester or the course. It gives an overall perspective. On the other hand, formative assessment provides continuous feedback at regular intervals. Thus, changes can be made before the final conclusions.

Formative assessment can be done in several ways. For instance, technology-enabled formative assessments are in vogue.

On January 27, 2020, the first case of Corona virus was reported in India. After just a few months, on March 25, an national lockdown of 21 days was implemented. Disasters have deadly impact on people's lives (Di Pietro, 2017). Even students had to face problems because of the pandemic's effect on education sector. Face-to-face (F2F) classes were not feasible due to requirement of social distancing, and the future of offline classes still seems to be uncertain. This led to a massive shift of education to online. Technological development has made online/distance education easy (McBrien et al., 2009). Online learning has its own advantages. Online education is easily accessible and can facilitate more visual teaching.

But one of the main concerns regarding online education is conducting and assessing assignments. Hannafin et al. (2003) noted that "the distant nature of web-based approaches renders difficult many observational and participatory assessments". Oncu and Cakir (2011) observed that informal assessment may be difficult for online instructors due to absence of F2F contact. Some papers have emphasized the importance of authentic assessments in online education (e.g., Kim et al., 2008; Robles & Braathen, 2002). Academic integrity should also be checked in online education (Kennedy et al., 2000; Simonson et al., 2006).Assessments can be used to promote academic self-regulation (Booth et al., 2003; Kim et al., 2008; Robles & Braathen, 2002).There is need to assess online discussion and collaboration too (Guangul et al., 2020; Meyer, 2006; Naismith, Lee, & Pilkington, 2011; Vonderwell et al., 2007). A student's performance in assignments and examinations reflects his/her understanding of the subject. They indicate the extent of success of student in reaching the desired outcomes. There are several ways to assess the progress of the student, which will be discussed later in this paper. Nagamani (2020) studied the use of Edmodo in encouraging the students to participate and collaborate especially in an online learning.

At the same time there should be emphasis on transparency of the criteria used for evaluation. It is important that a student should understand the feedback properly so that students can understand their shortcomings. Another major factor is ensuring academic honesty on the entire process. Lack of interaction between teacher and student can increase the ease for student to indulge in academic dishonesty, which undermines the spirit of education and should be addressed.

This chapter discusses summative assessment through home assignments and formative assessment through Quizziz, a web 2.0 tool.

9.2   OBJECTIVES OF THE STUDY

1. To understand problems of carrying out and assessing home assignments.
2. To figure out suitable methods of assessment with the help of responses given by students to the questionnaire prepared.

3. To give suggestions for making the process smoother.
4. To understand assessment through Quizziz.

9.3   REVIEW OF LITERATURE

9.3.1   What Is a Good Assessment and Feedback System?

Nicol and Macfarlane-Dick (2006: x) discovered seven principles of good feedback practice.

They opined that for students to understand the meaning of good performance, giving them only information about the criteria for evaluation is not enough, and there should be meetings and discussions specially to discuss the criteria for evaluation.

The following are the ten principles:

9.3.1.1   Clarify What Makes a Good Performance

Students should be clearly aware of desired outcomes and standards, goals, and criteria for assessment. (Students should be aware of what standards they should aim for.)

9.3.1.2   Encourage Students to Spend Time and Put Effort into Learning

Assessment tasks should encourage deep learning and proper understanding.

9.3.1.3   Give Students Information about Their Learning

Teachers are a crucial source of feedback for students. They are better able to judge students' progression to goals better than the students' peers or students themselves. With proper feedback students can evaluate their own shortcomings.

9.3.1.4   Encourage Positive Beliefs and Self-Esteem among Students

The system of assessment and feedback should encourage the learner to learn more. Motivation and self-esteem are important for students to increase their efforts to reach the goals and criteria.

9.3.1.5   Encourage Dialogue around Learning

Proper student–teacher and student–peer dialogue can make the learning process more effective. Some researchers have shown that student–teacher interaction is essential for the feedback to be effective. Feedback should not only involve the passing of information but the teacher should interact with students.

9.3.1.6   Facilitate the Development of Self-Assessment (Reflection) in Learning

Over the past decade there has been an uptrend in self-assessment in higher education. Self-assessment is an efficient way to increase students' learning. Teachers should create a proper structure and opportunities for students to do this.

9.3.1.7   Give Learners Choice in the Type of Assessment

There should be consultation with the students regarding their preferred way of evaluation, timing, etc.

9.3.1.8 Involve Students in Drafting Assessment Policy

Students should be involved in the process as it improves their understanding of assessment policy, and this would help them in their learning process.

9.3.1.9 Support the Development of Learning Communities

Assessment and feedback should enable the teacher to change their teaching style.

9.3.1.10 Help Teachers to Adapt and Change according to Students' Needs

The assessment and feedback process should provide adequate information to change the teacher's methodology.

9.4 RESEARCH METHODOLOGY

For research two questionnaires were prepared – one regarding assignments and the other regarding examination – and responses were gathered from 50 students who are currently studying Computer Science Engineering (I/II) in Geethanjali College of Engineering and Technology. In the questionnaire students were asked about the current method of assessment used in their institutions and their preferred method of assessment and evaluation and a few other questions.

9.5 ASSIGNMENTS

Assignments are given to students after the completion of a particular unit, and the assignment covers topics of that particular unit. Assignments aim to improve the understanding of students regarding the unit and improve students' problem-solving abilities. In the institution surveyed students are generally given objective type assignments, and subjective type questions which require students to write in detail about a topic.

Here students were asked which type of assignment they would prefer to do. They had following options given to them.

- **Solving numericals or questions**
  Problems are given to students that require them to use their knowledge of the topics and solve the given problem
- **Writing reports**
  Reports are documents that give information in an organized manner about a particular topic. To write an efficient report, students should have understanding of the topic in detail. So, these reports test and improve the students' understanding of the subject and their skill in organizing the information
- **Case studies**
  Case studies are critical analysis of a particular case. This improves students' understanding and tests their critical thinking.
- **Quizzes**
  In quizzes students are asked some questions under a specific time constraint. This can be made a team activity by making teams and treating it a sport.
- **Essay writing**
  In essay writing students are asked to write an essay regarding a specific topic.

- **Presentation**
  These can be done online and are a good reflection of students' understanding.
- **Fact sheet**
  Students can create a fact sheet on various topics or companies.

In the study students were asked which type of assignment they would prefer to do.
Analysis of questions of the questionnaire is presented through graphs:

Question 1: Which activities would you prefer to do as assignment?

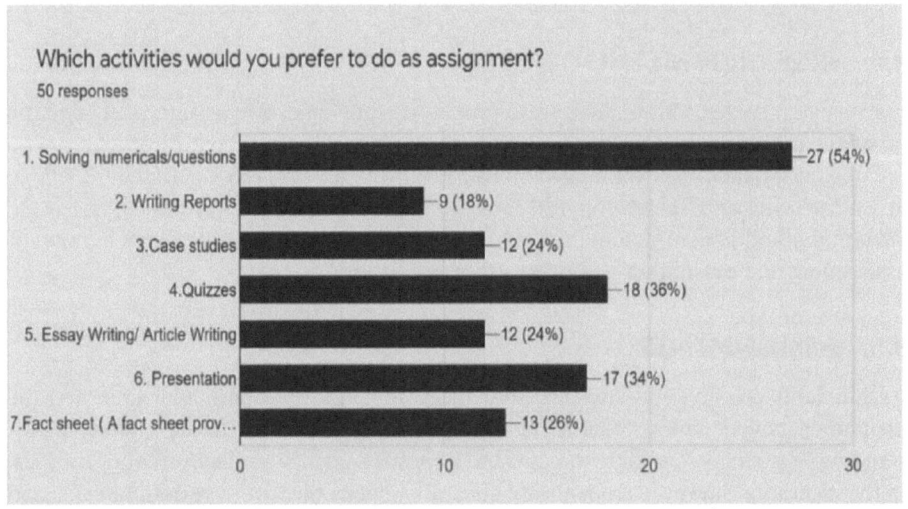

FIGURE 9.1 Graph showing the tasks that students would prefer to complete for an assignment.

Figure 9.1 shows that solving numericals or questions gained maximum popularity, followed by quizzes, and presentation got almost equal number of preferences.
Students were also asked whether they feel that assignments are monotonous.
Question 2: Do you feel assignments are monotonous?

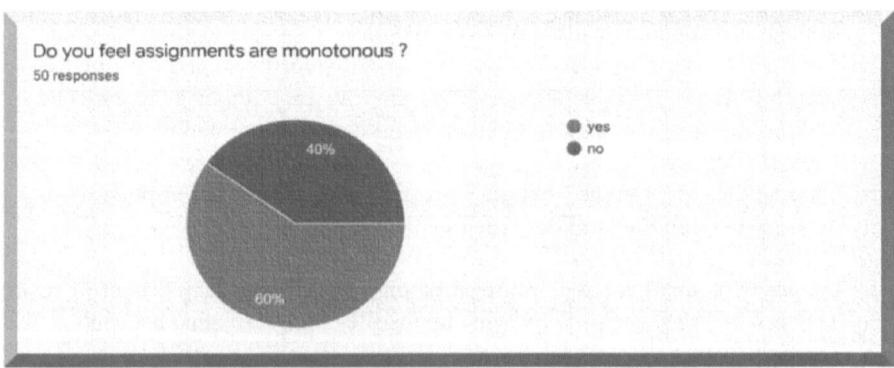

FIGURE 9.2 Pie Chart indicating whether or not the assignments are monotonous.

As you see in Figure 9.2, 60% of students feel that assignments given to them are monotonous. Many felt that the assignments require them to write a lot, so they become tedious after a while and they don't allow them to use their creativity.

Students were also asked about their preference for group activities. The majority felt that group activities can make assignments more interesting. Many believed that it would provide a great platform for interaction (Figure 9.3).

Question 3: Do you think group activities can make assignments more interesting?

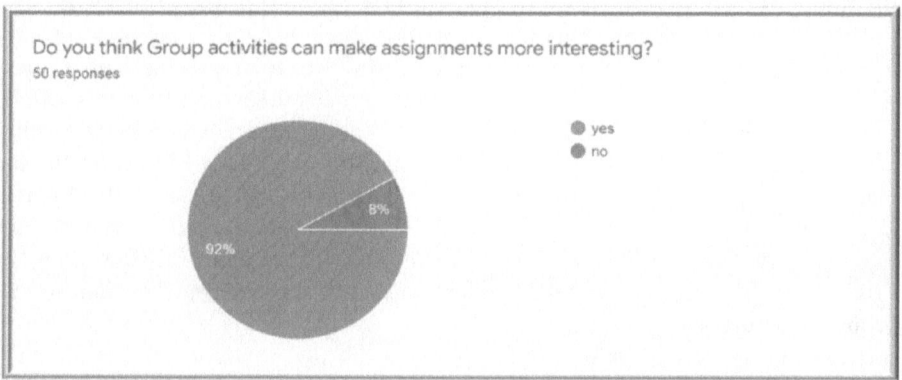

FIGURE 9.3 Pie chart illustrating if collaborative activities can make assignments more interesting.

After this, students were also asked about their view on peer/self-evaluation.

Question 4: Would you like your assignment reviewed by your peers/self or teachers?

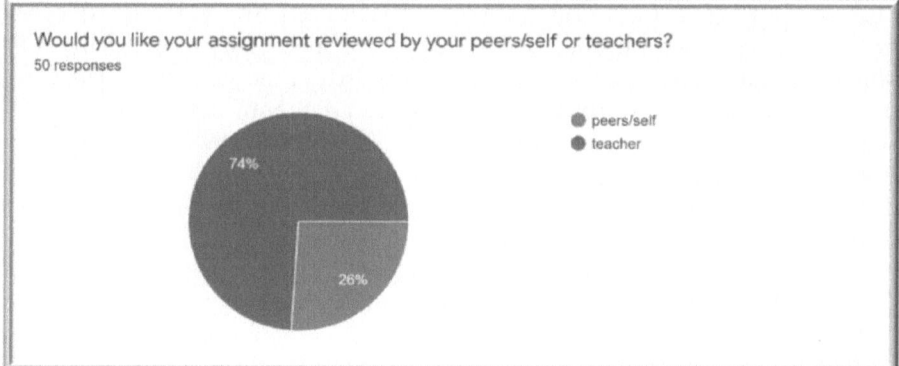

FIGURE 9.4 Pie chart indicating who among your teachers, peers, or you would want to review your assignment.

As seen in Figure 9.4, students preferred teacher's evaluation rather than peer evaluation. Many felt that the teacher has a better perception of the grading criteria, and they were unsure about their peers' judgement; for this reason they preferred evaluation by the teacher. So here we have understood that it is important to establish clear information about the goals, criteria, and standards and to win the confidence of students before implementing any such activities.

9.6   CHALLENGES AND SUGGESTIONS

The biggest challenge in this case is academic dishonesty. Students tend to copy the solutions from others, which would derail the purpose of assignments. By giving different questions to different students this challenge can be tackled. Another method of tackling this challenge is asking students to frame their own questions and answer them. The questions should reflect their understanding of the topic. Another challenge is assessment of the assignments. One single teacher is expected to evaluate about 60 assignments per class, which is hectic for the teacher. So, by introducing peer or self-evaluation this can be tackled. This can reduce the teacher's workload and at the same time improve students understanding (Yang & Tsai, 2010).But as we have seen earlier, students are reluctant for this as they feel teacher has better understanding of the criteria of evaluation. So proper structure should be made for this, and teacher should supervise the entire process. As mentioned before, many students felt that assignments are monotonous. So, the questions can be posed by audio or by online simulators to be more interesting. Students have given positive response regarding group activities. So, increasing group activities will improve students' engagement. It would be helpful if students and teachers mutually decide on which type of assignment is to be assigned for the particular unit.

9.7   EXAMINATIONS

Students have to take three examinations in a semester – two mid-term exams and one semester-end examination. One mid-term exam consists of first half of the semester syllabus and another covers the second half of the semester syllabus, and the semester-end exam covers the entire syllabus. Mid-term exams consist of 60% subjective questions and 40% objective questions, whereas semester exams consist of subjective questions with different assigned weights. Examinations are solely conducted to test students' understanding. During this pandemic open-book assessments were conducted in the institution studied.

In this questionnaire students were asked about their preference, and they could choose more than one option. The following options were given:

- **Proctored examination**
  Here, students write the exams and are proctored by invigilator or proctoring software. If the paper is subjective, then after the completion of the examination, students scan their answer sheet and upload it. If it is an objective paper, they can answer the questions on the given website itself (e.g., Moodle).

- **Open-book assessment**
  In this, students are given questions and they are free to refer to textbooks or any other book.
- **Professional presentation or demonstration or project**
  In this method, students prepare a project/presentation which reflects their understanding of the subject. The students can be evaluated with pre-defined criteria. This can even reflect students' creativity.
- **E- portfolio**
  Here, students represents their best work of the semester and give a formal paper on it.

The following were students' responses:

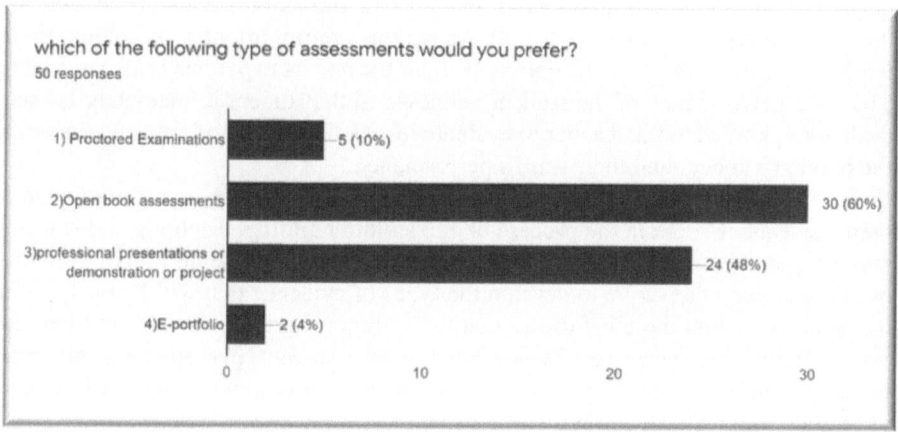

FIGURE 9.5   Graph illustrating the kinds of assignments you would prefer.

So, from Figure 9.5 it can be seen that majority of students prefer open-book assessment followed by a presentation.

9.8   CHALLENGES AND SUGGESTIONS

Similar to assignments, one of the biggest challenges in open-book assessment is academic dishonesty. Giving different questions to different students might reduce chances of cheating. By increasing the difficulty of the questions asked, there can be check on academic dishonesty. Another way is to reduce the examination time if the paper is of easy or moderate difficulty. By doing this, students just have enough time to answer and recheck their questions but not time for any other activities. It is advised that students should sign their institution's academic integrity intent before taking an examination. There is some evidences that suggest signing a pledge before taking an exam is an effective method to reduce academic dishonesty (McCabe et al., 2012).

9.8.1 QUIZZIZ

Quizziz is an interactive learning platform created primarily for understanding students' performance through multiple choice and or fill-in-the-blank questions. The easy-to-use templates are available on the website to create quizzes. In addition, there are a host of quizzes that are free resources that are tailored to subject and topic with varying difficulty levels. Teachers have to just check the relevance and before assigning the quiz. These quizzes can be live or assigned as homework.

At the end of the quiz, a leader board is displayed to inform who got the highest marks. Every question is time limited, and the quizzes can be instructor led or self-paced. The students' learning is augmented, modified, and redefined through this formative assessment tool.

Individual performance of the student in terms of time taken to answer the question and number of attempts is available in the automatically generated and downloaded report (Tables 9.1 to 9.4). As parents are important stakeholders in the teaching–learning process, the option to email the results to parents is also available. Thus, the performance of the student is known to the student immediately for self-evaluation; known to the teacher to evaluate instructional gaps, if any; and known to the parent to understand their ward's performance.

Formative evaluation, according to Bloom et al. (1971), is the application of comprehensive assessment in the process of curriculum building, teaching, and learning with the goal of enhancing any of these three processes. This means that in formative evaluation, one must strive to develop the types of evidence that will be most useful in the process, find the most useful method of reporting the evidence, and look for ways to reduce the negative effects related to assessment – perhaps by minimizing the judgmental aspects of evaluation or, at the very least, by having the formative evaluation users. (1971, p. 118).

In a major work that examined existing evidence on teacher evaluation Black and William (1998) stated that frequent formative assessment promoted student learning the most. They thought that students' capacity to honestly reflect on their own academic achievement and visualize their learning in the context of what they needed to improve was equally crucial. Formative assessment's major goal is to establish a link, or a relationship, between evaluation and learning. It primarily covers what

TABLE 9.1
Quiz Summary

Name	Value
Game Started On	Fri 28 Jan 2022, 02:45 PM
Game Type	Instructor-paced quiz
Participants	36
Total Attempts	36
Class Accuracy	54%
Game Ends On	Sat 29 Jan 2022,02:49 AM

TABLE 9.2
Participant Data

Rank	First Name	Attempt #	Accuracy	Score	Correct
1	20R11A0481	20	95%	17940	19
2	20R11A0487	20	90%	16990	18
3	20R11A0482	20	90%	16790	18
4	20R11A0499	20	80%	15390	16
5	20R11A0460	20	85%	15180	17
6	20R11A0456	20	80%	14740	16
7	20R11A0462	20	80%	14430	16
8	20R11A0495	19	75%	13940	15
9	20r11a0466	20	70%	12880	14
10	20r11a0471	19	75%	12860	15
11	20R11A0490	20	75%	12680	15
12	20R11A0489	19	70%	12660	14
13	20r11a0450	20	70%	12190	14
14	20R11A0453	20	70%	12050	14
15	20r11a0470	20	65%	11760	13
16	20R11A0475	20	70%	11760	14
17	20R11A0486	19	65%	11710	13
18	20R11A0494	20	65%	11650	13
19	20R11A0476	20	60%	11060	12
20	20R11A0465	18	55%	10430	11
21	20R11A0463	19	55%	10390	11
22	20R11A0464	20	65%	10300	13
23	20R11A0483	20	60%	9940	12
24	20r11a0461	20	50%	9400	10
25	21R15A0409	20	50%	8810	10
26	20R11A0478	20	45%	7780	9
27	20R11A0457	14	40%	7510	8
28	20R11A0491	19	40%	7430	8
29	20r11a0455	17	45%	6280	9
30	20R11A0480	3	15%	2850	3
31	20R11A0468	2	5%	890	1
32	21R15A0410	3	5%	820	1
33	21R15A0406	0	0%	0	0

occurs in the classroom, the type of interactions between and/or among instructors and students, and the quality of the educational experience.

9.9 CONCLUSION

The Covid-19 epidemic, according to the World Economic Forum, has redefined how people learn and teach. We may need to introduce some much-needed change in order to find novel answers to our challenges. Traditional classrooms

TABLE 9.3

Time Taken to Answer the Questions

#	Question	Question Type	Question Accuracy	Average Time per Question (mm:ss)	Correct
1	Brisk	Multiple Choice	30%	00:14	11
2	Speedy	Multiple Choice	52%	00:10	19
3	Handsome	Multiple Choice	52%	00:09	19
4	Rapid	Multiple Choice	47%	00:14	17
5	Quick	Multiple Choice	58%	00:09	21
6	Swift	Multiple Choice	25%	00:14	9
7	Lay	Multiple Choice	19%	00:14	7
8	Earth	Multiple Choice	77%	00:07	28
9	Freedom	Multiple Choice	72%	00:05	26
10	Table	Multiple Choice	44%	00:09	16
11	Money	Multiple Choice	66%	00:08	24
12	Weather	Multiple Choice	58%	00:08	21
13	Key	Multiple Choice	30%	00:12	11
14	Book	Multiple Choice	33%	00:08	12
15	Rubber	Multiple Choice	50%	00:11	18
16	Beautiful	Multiple Choice	80%	00:07	29
17	Fast	Multiple Choice	75%	00:06	27
18	Home	Multiple Choice	55%	00:08	20
19	Cooling	Multiple Choice	72%	00:07	26
20	Home	Multiple Choice	86%	00:05	31
			54%	**03:05**	**392**

have been habitual for both teachers and students, yet we have no choice but to adjust to this shift. It has the potential to deliver a slew of unexpected improvements to the field of education. Education is a guiding force for development, and whatever happens, education should not stop. This pandemic has taught us the importance of adapting. Things are unpredictable, and adapting to the required changes is the only way. In these situations, both students and teachers should have understanding of each other's limitations. It is important to increase students' engagement in the process, especially now, because of lack of accessibility and reduced interaction between teachers and students because of the pandemic. If students are involved in decision-making about assessment policy, this would help them to have better understanding of the evaluation criteria and will even increase their engagement in the process. Proper opportunities should be given to the students to self-evaluate or evaluate their peers. But proper structure and supervision is needed for this. Students should be also encouraged to have discussions with their peers around learning, and teachers should encourage formation of learning communities.

While providing online education, an educator must have five things in mind: instruction, material, motivation, connections, and mental wellness (Martin, 2020).

TABLE 9.4
Overview

#	Question	Question Type	Question Accuracy	Average Time per Question (mm:ss)	Correct	Incorrect	Unattempted	20R11A0481 (20R11A0481)	20R11A0487 (20R11A0487)	20R11A0482 (20R11A0482)
1	Brisk	Multiple Choice	30%	00:14	11	10	15	Walk	Walk	Walk
2	Speedy	Multiple Choice	52%	00:10	19	8	9	Recovery	Recovery	Recovery
3	Handsome	Multiple Choice	52%	00:09	19	9	8	Boy	Gentlemen	Boy
4	Rapid	Multiple Choice	47%	00:14	17	11	8	Progress	Progress	Progress
5	Quick	Multiple Choice	58%	00:09	21	8	8	Glance	Glance	Glance
6	Swift	Multiple Choice	25%	00:14	9	19	8	Race	Action	Race
7	Lay	Multiple Choice	19%	00:14	7	22	7	Emphasis	Emphasis	Emphasis
8	Earth	Multiple Choice	77%	00:07	28	1	7	Quake	Quake	Quake
9	Freedom	Multiple Choice	72%	00:05	26	3	7	Fighter	Fighter	Fighter
10	Table	Multiple Choice	44%	00:09	16	13	7	Salt	Salt	Salt
11	Money	Multiple Choice	66%	00:08	24	5	7	Lender	Lender	Lender
12	Weather	Multiple Choice	58%	00:08	21	8	7	Forecast	Forecast	Forecast
13	Key	Multiple Choice	30%	00:12	11	18	7	Bunch	Bunch	Group
14	Book	Multiple Choice	33%	00:08	12	17	7	Worm	Worm	Worm
15	Rubber	Multiple Choice	50%	00:11	18	11	7	Stamp	Stamp	Stamp
16	Beautiful	Multiple Choice	80%	00:07	29	0	7	Girl	Girl	Girl
17	Fast	Multiple Choice	75%	00:06	27	2	7	Food	Food	Food
18	Home	Multiple Choice	55%	00:08	20	10	6	Maker	Maker	Maker
19	Cooling	Multiple Choice	72%	00:07	26	5	5	Glass	Lens	Glass
20	Home	Multiple Choice	86%	00:05	31	0	5	Work	Work	Work
			54%	**03:05**	**392**	**180**	**148**	**95%**	**90%**	**90%**

So, teachers should try to encourage positive motivational beliefs and self-esteem in the students. Management should also seek feedback of students regarding teacher performance and how can they improve. Ultimately, students and teachers should collaborate, and mutual understanding should make the process smoother and minimize the pandemic's negative effect on the education sector.

REFERENCES

Black, P., & William D. (1998). Assessment in classroom learning. Assessment in Education, 5(1), 7–74. doi:10.1080/0969595980050102

Bloom, B.S., Madaus, G.F., & Hastings, J.T. (1971). Handbook on Formative and Summative Evaluation of Student Learning. New York: McGraw-Hill.

Booth, A., Johnson, D., Granger, D.A., Crouter, A.C., & McHale, S. (2003). Testosterone and child and adolescent adjustment: The moderating role of parent-child relationships. Developmental Psychology, 39(1), 85–98. doi:10.1037/0012-1649.39.1.85

Di Pietro, G. (2017). The academic impact of natural disasters: Evidence from the L'Aquila earthquake. Education Economics, 26(1), 62–77. https://doi.org/10.1080/09645292.2017.1394984

Guangul, F.M., Suhail, A.H., Khalit, M.I. et al. (2020). Challenges of remote assessment in higher education in the context of COVID-19: A case study of Middle East College. Educational Assessment, Evaluation and Accountability, 32, 519–535. https://doi.org/10.1007/s11092-020-09340-w

Hannafin, M., Oliver, K., Hill, J.R., Glazer, E., & Sharma, P. (2003). Cognitive and learning factors in web-based distance learning environments. In M. G. Moore & W. G. Anderson (Eds.), Handbook of Distance Education (pp. 245–260). Mahwah, NJ: Erlbaum.

Huba, M.E., & Freed, J.E. (2000). Learner-Centered Assessment on College Campuses: Sifting the Focus from Teaching to Learning. Community College Journal of Research and Practice, 24.

Kennedy, K., Nowak, S., Raghuraman, R., Thomas, J., & Davis, S.F. (2000). Academic dishonesty and distance learning: Student and faculty views. College Student Journal, 34(2), 309–314.

Kim, N., Smith, M.J., & Maeng, K. (2008). Assessment in online distance education: A comparison of three online programs at a university. Online Journal of Distance Learning Administration, 11(1). Retrieved from http://www.westga.edu/~distance/ojdla/spring111/kim111.html

Lambert, D., & Lines, D. (2000). Understanding Assessment: Purposes, Perceptions, Practice, Routledge/Falmer, 2000 - Education.

Martin, A. (2020). How to optimize online learning in the age of coronavirus (COVID-19): A 5-point guide for educators. UNSW Newsroom, 53(9), 1–30.

McBrien, J.L., Cheng, R., & Jones, P. (2009). Virtual spaces: Employing a synchronous online classroom to facilitate student engagement in online learning. The International Review of Research in Open and Distributed Learning, 10(3), 1–17.

McCabe, D., Butterfield, K., & Trevino, L. (2012). Cheating in college: Why students do it and what 633 educators can do about it. Management and Organization, Baltimore: Johns Hopkins Press.

Meyer, K.A. (2006). The method (and madness) of evaluating online discussions. Journal of Asynchronous Learning Networks, 10(4), 83–97. Retrieved from http://www.sloanconsortium.org/sites/default/files/v10n4_meyer1_0.pdf

Nagamani, B. (2020). Flip-Flop with ICT. English Classroom, Bi-Annual Journal published by Regional Institute of English, South India, ISSN No: 2250-2831, Volume 22.

Naismith, L., Lee, B.-H., & Pilkington, R. (2011). Collaborative learning with a wiki: Differences in perceived usefulness in two contexts of use. Journal of Computer Assisted Learning, 27(3), 228 –242. doi:10.1111/j.1365-2729.2010.00393.x

Nicol, D., & Macfarlane-Dick, D. (2006). Formative assessment and self-regulated learning: A model and seven principles of good feedback practice. Studies in Higher Education, 31(2), 199–218.

Oncu, S., & Cakir, H. (2011). Research in online learning environments: Priorities and methodologies. Computers & Education, 57(1), 1098–1108. doi:10.1016/j.compedu.2010.12.009

Palomba, C., & Banta, T.W. (1999). Assessment Essentials: Planning, Implementing, and Improving Assessment in Higher Education. San Francisco: Jossey-Bass, Inc.

Robles, M., & Braathen, S. (2002). Online assessment techniques. Delta Pi Epsilon Journal, 44(1), 39–49.

Simonson, M., Smaldino, S.E., Albright, M., & Zvacek, S. (2006). Teaching and learning at a distance: Foundations of distance education (3rd ed.). Upper Saddle River, NJ: Merrill/Prentice Hall.

Vonderwell, S., Liang, X., Alderman, K. (2007). Asynchronous discussions and assessment in online learning. Journal of Research on Technology in Education, 39(3), 309–328.

Yang, Y.F., & Tsai, C.-C. (2010). Conceptions of and approaches to learning through online peer assessment. Learning and Instruction, 20(1), 72–83. doi:10.1016/j.learninstruc.2009.01.003

10 Curriculum Mapping for Attainment of Student Outcomes from the Perspectives of Senior-Year Students and Instructors

Ebru Dulekgurgen, Cigdem Yangin-Gomec, and Didem Okutman Tas
Istanbul Technical University
Istanbul, Turkey

CONTENTS

DOI: 10.1201/9781003083160-14

10.1   INTRODUCTION: BACKGROUND AND DRIVING FORCES

Accreditation, the process of quality control and assurance in higher education, is recognized as a powerful practice in increasing institutional awareness of quality (Sin et al., 2017) and effectively contributing to continuous improvement of education and its quality in a broader sense. As it confirms the active process of continuous collective and individual efforts by the higher education institutions at various levels for achievement of the required criteria, accreditation represents a quality guarantee for a program and its graduates: today's students and tomorrow's professionals. The two pillars of quality assurance and accreditation might be addressed as keeping track of the level of attainment of the "student/learning outcomes" as the benchmark and "continuous quality improvement" as the core process (Dulekgurgen et al., 2019a). Assessment of student outcomes (SOs) might be described as "what students are expected to know and be able to do by the time of graduation ... [and] relate to the knowledge, skills, and behaviors that students acquire as they progress through the program" (ABET EAC 2016; 2018; 2022), and are therefore of particular importance. Accordingly, in the scope of periodical reviews for accreditation/re-accreditation, it is necessary to document the processes for regularly assessing and evaluating the extent to which the SOs are being attained. Moreover, it is also necessary to describe how the results of those processes are utilized in supporting continuous improvement of the programs.

Assessment and evaluation in line with the continuous quality improvement process have been receiving significant attention on the higher educational agenda. In terms of assessment and evaluation (A&E) of the level of attainment of the SOs, the context of data collection might include, but is not limited to, specific exam questions, student portfolios, internally developed assessment exams, senior-year design projects, nationally normed exams, oral exams, focus groups, industrial advisory committee meetings, or other tools and means of assessment that are relevant and appropriate to the program.

Using analytical rubrics provides valuable data for direct assessment of the level of SO achievement, and thus has been practiced widely as one of the fundamental and best practices in direct SO A&E (Dulekgurgen et al., 2018; Felder and Brent, 2003; Trevisan et al., 1999). Carefully prepared rubrics may lead to more objective assessment of student performance, thus promoting academic standards (Chan and Ho, 2019). Some other tools and methods for assessing students' performances with regard to the SOs are surveys (i.e., applied to students, alumni, employers, industrial advisory board members, etc.), exit interviews targeting graduating senior students, performance on standardized tests (e.g., fundamentals of engineering (FE) exam, GRE), and job placement data of graduates (Felder and Brent, 2003; Scales et al., 1998).

In addition to some other aspects of education and learning (e.g. summer internships, extracurricular activities related to students' academic achievements, self-learning activities), achievement of SOs can be reached mainly by learning in well-designed courses that promote multiple learning outcomes or through enrolling in numerous courses that focus on different aspects of learning (Mintz and Tal, 2014). Accordingly, keeping track of achievement of SOs by a particular cohort, or in a broader sense, by all students enrolled in a program, requires a comprehensive and well-structured SO A&E plan. In this context, a curriculum map overarching all four years of an undergraduate program becomes a valuable and fundamental tool for SO A&E at the program level. In a broad sense, curriculum mapping might be described as a method for relating the instructional activities and the learning/student outcomes, together with the means of assessing to which extent those outcomes are being attained. Mostly given in a matrix format, curriculum maps provide standardization and visualization of those relations (UDel CTAL, 2020). Curriculum maps might be prepared

- *at different levels*: for a single course by single/multiple instructors, for a whole curriculum by the entire faculty,
- *at varying complexity and details*: activity schedule, content and covered topics, targeted learning/student outcomes, assessment methods, detailed list of assessment tools, clearly described performance indicators, etc.,
- *for different purposes*: course- or program-level standardization, communication, assessment, improvement.

In the context of quality assurance in higher education and meeting accreditation criteria, a curriculum map prepared by the joint effort and involvement of the entire faculty of a four-year undergraduate program might be addressed as one of the fundamental roadmaps of the SO A&E process at program level (Dulekgurgen et al., 2019b).

As mentioned above, student portfolios (SPs) might be considered an alternative tool in the SO A&E process and also as a powerful self-regulated learning approach for the students (Lam, 2014). Preparing portfolios makes the learning process more effective and visible, and they are evidence of knowledge, skills, and attitudes (Bader et al., 2019). Moreover, since SOs are expected to guide students' learning, it is also important to see the perceptions of students who are the recipients of education and the active actors of the learning process (Kumpas-Lenk et al., 2018). Student portfolios might include student work, surveys, sample exams, homework, etc. Accordingly, they are listed as possible assessment tools that might be utilized in course- and program-level assessment (Felder and Brent, 2003). However, it is not recommended to use them as the primary tool and data source of assessment at the program level, and issues related to validity and reliability while processing the student portfolios for assessment need to be addressed (Gredler, 1995).

In particular, student portfolios are reported as lacking a systematic process for SO assessment (ElAli, 2018). Although SPs have been promoted and used for engineering program assessment in previous studies, it is reported that using conventional (hard-copy) portfolios to assess attainment level of SOs would be time-consuming and hard to manage (Felder and Brent, 2003). Moreover, there are also some observed drawbacks associated with student portfolios, namely,

- need for a representative number of students to collect materials,
- students including only their best work,
- difficulty in assessing SOs without extensive examination of a large number of portfolios, etc. (Morgan et al., 2000).

Considering both disadvantages and benefits mentioned above, student portfolios might still serve as a valuable source of information, especially reflecting students' perceptions and viewpoints.

With the latter as the starting point, the study presented here was conducted at the Environmental Engineering Department of Istanbul Technical University (ITU), a public university hosting more than 25,000 undergraduate students at the time of the study and being one of the top-ranked universities in Turkey. Quality assurance in higher education, particularly in the field of engineering, is among the top priorities of the institute, which holds the leading position in the country in terms of the total number of internationally accredited programs: 25 engineering undergraduate programs are accredited by the ABET EAC (Engineering Accreditation Commission), which is the flag-ship commission of the non-profit, non-governmental organization that reviews and accredits engineering programs worldwide (ABET, 2022). The Environmental Engineering Undergraduate Program (EEUP) at ITU is among those programs accredited by the ABET EAC (ABET, 2022; ITU EED, 2022).

Several studies on quality assurance and accreditation-related works in a variety of engineering undergraduate programs are accessible in the literature (Aoudia and Abu-Alqahsi, 2015; Christy and Lima, 1998; ElAli, 2018; Felder and Brent, 2003; Scales et al., 1998; Shafi et al., 2019; Soeiro et al., 2009; Trevisan et al., 1999). However, to our knowledge, those conducted in environmental engineering undergraduate programs are relatively limited in number (Mintz and Tal, 2014). Accordingly, the study presented here is considered to have a useful contribution to the knowledge build-up in this area. Based on these observations, results from the literature, and the works carried out at the ITU Environmental Engineering Department for quality assurance and continuous improvement of the higher education delivered in the engineering undergraduate program. The driving force for this study was to provide answers to the following targeted research questions:

> **Research question 1:** *The curriculum map utilized in the previous accreditation review cycle of the environmental engineering program was structured mainly by using the inputs of the instructors, yet the students are at the center of higher education as the primary constituent. Accordingly, how did the students perceive and reflect on the relations between the courses they have taken and contribution of those courses to attainment of the targeted student outcomes?*
>
> **Research question 2:** *Is it possible to utilize the students' perceptions/ reflections presented in their student portfolio as inputs to further fine-tune the curriculum map and/or as inputs for assessment and evaluation of level of achievement of the targeted student outcomes to serve for continuous improvement of the education provided through the accredited environmental engineering program?*

Accordingly, the intend of this study was not to utilize and process student portfolios directly in the SO A&E but to showcase the similarities and differences between students' perception and instructors' views regarding curriculum mapping utilized for assessing the attainment of the student outcomes, which is the ABET EAC's legacy 11 SOs (ABET Webinar#7, 2020), overarching the four-year environmental engineering curriculum. For that, the senior-year students of the Environmental Engineering Undergraduate Program (EEUP), who were preparing their senior-year capstone design projects – their graduation engineering design projects (GDPs) in teams – were invited to join the study on a voluntary basis. One team of three students, was asked to prepare their own curriculum map relating the compulsory courses of the program to the SOs in contributing to acquiring and/or improving their knowledge, skills, and behaviors as they progressed through the program and to include that in their combined student (team) portfolio. Particularly, the curriculum map prepared by the students was compared to that constructed by the instructors to determine the similarities and differences. Students' own perception and comments on why they matched particular SOs with certain courses were exemplified in details through two selected courses. Possible reasons, especially for the differences, were also discussed. Based on the analysis of the presented results, some recommendations for future improvements in preparation and evaluation of student portfolios, and for their possible function in the SO A&E plan are offered.

10.2 METHODOLOGICAL APPROACH

10.2.1 Procedure and Participants

As stated above, the study was conducted at the ITU Environmental Engineering Department (ITU EED, 2022). The instructor-based curriculum map for SO A&E utilized in this study was constructed by the involvement and collective effort of all 45 faculty members of the department and fine-tuned by the related committees of the department (i.e., ITU EED Accreditation Coordination Committee [ACC] and the Department's Curriculum Development Committee [DCDC]) in a series of academic board meetings.

For appraisal of the research questions of this study, senior-year students who teamed up for conducting their graduation engineering design projects were invited to participate on a voluntary basis, and one team consisting of three students responded positively to that call and provided informed consent. Note that formation of the "student teams" for the senior-year capstone engineering design projects was initially organized by the related coordinators in the department, based on certain criteria (e.g., students' GPAs, equal gender distribution), which provided a balanced student profile distribution in each team. Apparently the number of students who volunteered to join this study was limited to be representative of the entire senior-year cohort. Also, the small student population size had the potential risk of self-selection bias during their contributions. Yet, the academic profiles of the members of the voluntary team were balanced as indicated by their GPAs being representative of the medium and maximum GPAs of that year's senior cohort. Accordingly, it was decided to conduct the study with the available student population as an initial effort to facilitate similar implementations in future.

10.2.2　Student Outcomes

The student outcomes (SOs), which are the subject matter of this study, are listed below. Those 11 SOs are the ones described by the EAC of ABET, a USA-based international quality assurance and accreditation agency. They were used first in the EC2000 documents (Engineering Criteria, 2000) and were in use as one of the benchmarks in assessment, evaluation, and quality assurance in engineering education until recently. The data presented in this study were collected from the ITU Environmental Engineering Undergraduate Program, which is an ABET EAC–accredited program (ABET, 2022); therefore the ABET EAC's legacy 11 SOs, also known as SO a-to-k (ABET Webinar, 2020), were adapted and utilized by the program at the time of the study as the main targets to be achieved through the four-year curriculum and the primary assets of the education and learning experiences offered to the students. Note that, after two decades of utilization, those 11 legacy SOs have recently been revised and a new set of seven SOs were adopted as of the 2019–20 accreditation review cycle (ABET EAC, 2018; ABET Webinar, 2020). However, at the time of this study, the former 11 SOs were still in effect (ABET EAC, 2016) and thus were implemented by the program. The curriculum maps developed by the instructors and the students presented here were targeting those 11 legacy SOs (the letters a to k in parentheses correspond to the original notation used by ABET).

> *SO1 An ability to apply knowledge of mathematics, science, and engineering (a)*
> *SO2 An ability to design and conduct experiments, as well as to analyze and interpret data (b)*
> *SO3 An ability to design a system, component, or process to meet desired needs within realistic constraints such as economic, environmental, social, political, ethical, health and safety, manufacturability, and sustainability (c)*
> *SO4 An ability to function on multidisciplinary teams (d)*
> *SO5 An ability to identify, formulate, and solve engineering problems (e)*
> *SO6 An understanding of professional and ethical responsibility (f)*
> *SO7 An ability to communicate effectively (g)*
> *SO8 The broad education necessary to understand the impact of engineering solutions in a global, economic, environmental, and societal context (h)*
> *SO9 A recognition of the need for, and an ability to engage in lifelong learning (i)*
> *SO10 A knowledge of contemporary issues (j)*
> *SO11 An ability to use the techniques, skills, and modern engineering tools necessary for engineering practice (k)*

10.2.3　Student Portfolio

Three senior-year student volunteers were asked to present their student portfolio and include their own version of the curriculum map for the ITU Environmental Engineering Undergraduate Program. After a brief description of the minimum requirements of a portfolio and a curriculum map, they were left free to decide what to include in their portfolio and to what extent to present their work and comprehension. Accordingly, the students prepared their own version of the matrix relating the courses to the SOs (the student-based curriculum mapping), included their own comments on why they did such particular matchings, and provided samples from the works they completed in the courses (homework assignments, term projects, laboratory reports, etc.) all together forming the sample self-organized student portfolio.

10.2.4 CURRICULUM MAPPING FOR STUDENT OUTCOME ASSESSMENT AND EVALUATION

Each course offered by ITU has a four-page, bilingual identification document (native language/English), called Course Catalog/Description Form (CCF), which describes all fundamental features of the course. The CCFs are prepared and updated by the instructors of the courses and are available online to the internal constituents of the institute (i.e., students, faculty members, administrators, etc.). In addition to the basic technicalities, such as the name, code, semester, year, local credits, European Credit Transfer System (ECTS) credits, type (i.e., compulsory, elective), prerequisites, etc., of the course, each CCF also includes the following features: description, objectives, and learning outcomes of the course; assessment criteria (type and number of activities, percentage of final grade); references; homework and/or project assignments; course plan listing the topics covered at each week and relating those to the course learning outcomes; and the matrix showing the relation between the course and its level of contribution to attainment of the ABET EAC's student outcomes.

In order to construct the curriculum map covering the four-year curriculum of the program and be able to use it for SO A&E at the program level, it was necessary to ensure the participation of the entire faculty. Accordingly, all faculty members of the department were asked to reevaluate the CCFs of their related courses individually or in teams. The initial feedback provided by the instructors was then processed by the accreditation coordination and curriculum development committees of the department, leading and coordinating the mapping process. For an effective, relevant, and functional SO A&E, only compulsory courses were included in the curriculum map. The compulsory courses were listed in the map in a sequential order, from freshman to senior year, and the level of contribution of each course provided by the instructors was marked for each of the 11 SOs, which were also listed in the matrix.

The resulting initial curriculum map was shared with the faculty members and discussed in a number of academic board meetings for verification and/or further improvement. Communication and joint discussions enabled minimizing/eliminating the gaps in the map, reviewing and revising some courses (when applicable) in terms of contribution to attainment of the SOs, and fine-tuning the curriculum map. The final version of the map was communicated to all relevant parties (i.e., instructors, related committees, outcome coordinators in charge of advanced SO A&E, etc.) and implemented in the SO A&E plan of the program. The visual presentation of the continuous improvement strategy of the ITU Environmental Engineering undergraduate program, including the assessment and evaluation process flow, can be found in the department's publically accessible web-page (ITU EED, 2022).

A simplified version of the curriculum map prepared by direct involvement of the course instructors is given in Table 10.1. Note that the original map included all three levels of relative contribution of the compulsory courses assigned by the faculty members: [level-1] Introduced; [level-2] Reinforced; [level-3] Emphasized, Assessed, and Evaluated (data not shown here). However, since SO A&E by the instructors and the outcome coordinators were carried out by collecting assessment data from the compulsory courses with the highest level of contribution (level-3), the map presented in Table 10.1 was simplified accordingly. The volunteer senior-year students prepared their own curriculum map in a similar format and presented that in their sample student (team) portfolio. The students' perspective of relating the compulsory courses to achievement of the SOs is also included in the matrix seen in Table 10.1.

TABLE 10.1

Simplified Version of the ITU EEUP's Curriculum Map[a] Relating the Compulsory Courses to the SOs Contributing to Their Attainment: (I) Provided by the Instructors and (S) Marked by the Voluntary Students

Compulsory Courses[b]	Term[b]	Student Outcomes										
		SO1	SO2	SO3	SO4[c]	SO5	SO6	SO7	SO8	SO9	SO10	SO11
1st year – Freshman												
113/E Intro to Environmental Eng	F	S								S	S	
2nd year – Sophomore												
211/E Environmental Chemistry I	F	I	I+S							S	S	
213/E Environmental Microbiology	F	S	I+S								S	
242/E Urban Hydrology	F	I+S								S	S	
212/E Environmental Chemistry II	Sp	I	I+S							S	S	
228/E Earth Science	Sp	S										
3rd year – Junior												
327/E Chemical Processes	F	I				S						
329/E Solid Waste Management	F	S	I	S		I		S		S	S	
343/E Water Quality Management	F					I+S	S	S				
345/E Control and Auto in Environ Facilities	F			S								I
347/E Unit Operations	F	I+S		S		I+S						

(Continued)

TABLE 10.1 *(Continued)*
Simplified Version of the ITU EEUP's Curriculum Map[a] Relating the Compulsory Courses to the SOs Contributing to Their Attainment: (I) Provided by the Instructors and (S) Marked by the Voluntary Students

Compulsory Courses[b]	Term[b]	Student Outcomes										
		SO1	SO2	SO3	SO4[c]	SO5	SO6	SO7	SO8	SO9	SO10	SO11
320/E Biological Processes	Sp	I+S		S		I+S						
324/E Unit Operations and Processes Lab	Sp	I	I+S				S					
326/E Air Pollution	Sp		I+S	S		I+S	S	S		S	S	
328/E Water Supply and Wastewater Disposal	Sp	S		I+S		S		S				I
330/E Water Treatment Plant Design	Sp	S		I+S		S	S	I+S				I
4th year - Senior												
427/E Environ Modelling Principles	F	S		S		I+S	S	S		S	S	I+S
429/E Industrial Pollution Control	F					I+S		I+S				
433/E Hazard. Waste Management	F								I	S	I	
437/E Wastewater Treat. Plant Design	F	S		I+S		S	S	I+S				I
471/E Ethics in Environmental Eng	F						I+S	I+S				
432/E Environmental Economics	Sp								I	S		I
442/E Environmental Law	Sp									I+S	I	
456/E Occupational Health and Safety	Sp									S	I	
492/E Graduation Design Project	F/Sp	I+S		I+S	I	I+S		I+S	I+S	S	S	I+S

[a] For the ITU EEUP curriculum valid between Fall 2011 and 2017.

[b] Compulsory courses offered by the faculty members at fall (F) and spring (Sp) terms: based on instructors' map and only those with highest level of contribution are listed ([level-3] emphasized, assessed, and evaluated by the instructors and the SO Coordinators).

[c] Students related the attainment of this outcome to some courses other than those listed here; please see text for details.

10.3 RESULTS AND DISCUSSIONS

10.3.1 COMPARATIVE EVALUATION OF CURRICULUM MAPPING

When performed across a four-year program, curriculum mapping provides the pattern and logic of curriculum design and a snapshot of the scholarly advancement of the students through the curriculum; hence it has the potential of showing the points that require reinforcement and improvement in terms of teaching and learning at program level (UDel CTAL, 2020). A simplified version of the curriculum map (including only compulsory courses with the highest contribution), developed by direct involvement of the instructors of the ITU Environmental Engineering Undergraduate Program and utilized for assessment and evaluation of the ABET EAC's legacy 11 SOs at program level, is given in Table 10.1, where entries provided by the instructors are indicated by "I". For side-by-side comparison, the compulsory courses specified by the senior-year capstone design project student volunteer team as those contributed to their knowledge, skills, etc., are marked on the same table as "S", and those marked by both constituents are shown as "I+S".

As apparent from Table 10.1, some aspects of the curriculum map presented by the students in their student portfolio matched those suggested by the instructors. Yet, there were some differences in students' opinions, generally in the direction of relating more courses to the SOs in comparison to those mapped by the instructors, which reflects a broader student perception. Most notable differences in this context were obtained for "SO6 – an understanding of professional and ethical responsibility", "SO9 – a recognition of the need for, and an ability to engage in life-long learning", and "SO10 – a knowledge of contemporary issues". For instance, according to the instructor-based mapping, courses contributing at the highest level to attainment of SO6 and SO9 were "Ethics in Environmental Engineering" and "Environmental Law", respectively, whereas the students marked those and six and eleven more courses, respectively, as contributing to achievement of those outcomes. Among those, freshman, sophomore, junior, and senior year courses were listed by them (e.g., "Intro to Environmental Engineering", "Air Pollution", "Environmental Economics").

Aside from the possible partial bias introduced by the way the students prepared their map without differentiating between the three different levels of contribution of the courses the way that is described above, some of those additional matchings done by the students were assumed to be implying the possible points of "students' self-learning", and were considered as pointing to the strengths of the curriculum from students' perspective (UDel CTAL, 2020). For SO10, three senior-year courses, namely "Hazardous Waste Management", "Environmental Law", and "Occupational Health and Safety", were suggested by the instructors as contributing to attainment of that outcome at the highest level. Instead, nine other courses from freshman, sophomore, junior, and senior years were correlated to the same outcome by the students.

One possible reason for the differences between the instructor- and student-based mapping was identified as follows: While preparing their map and portfolio, the students considered the courses they took (in particular, the compulsory courses) and mainly the assignments they completed (i.e., lab reports, term-papers, engineering design projects, presentations) as relevant inputs that improved their knowledge, skills, and behaviors. However, they did not address the formal exam-based assessments (i.e., relevant questions/sections in the midterm exams, quizzes, final exams, etc.) in their student portfolio as direct tools contributing to achievement of the SOs. On the other hand, in addition to using student work in SO assessment, not the overall grades but the partial scores from some exam questions, particularly targeting the SOs, were included by many instructors as one of the inputs while assessing the level of achievement of the targeted SOs. Another reason for the differences between the two maps was that the students did not provide a scale or specify the level of contribution (e.g., low, partly, high) when relating the courses to the SOs in their curriculum map. The latter was interpreted as an assessment artifact and a possible source of partial bias when comparing the two maps.

Similarities and differences between the instructor- and student-based curriculum mapping are visually summarized in Figure 10.1, which clearly shows that the students projected a higher number of courses than the instructors for SO1, SO3, SO6, SO9, and SO10. The number of matching courses, marked both by the instructors and by the students (I+S), was high for SO2 and SO5, whereas matching was below 50% for SO1, SO3, SO6, SO8, SO9 and SO11.

As mentioned above, the courses marked by the students and the instructors as contributing to attainment of SO10 were different from each other. Also, comparison

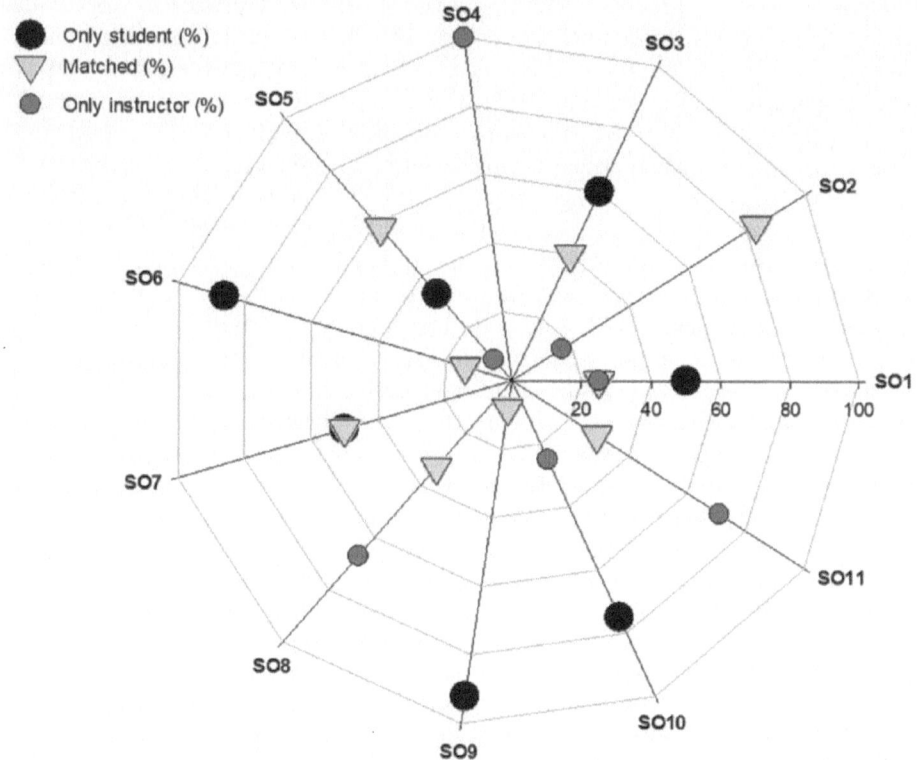

FIGURE 10.1 Comparison of curriculum mapping by the instructors and the voluntary senior-year students for attainment of the 11 legacy SOs of ABET EAC.

of the curriculum maps reflected that the instructors marked the senior-year capstone engineering design course (CEV492/E Graduation Design Project) as contributing to attainment of SO4, but the students did not relate any course listed in Table 10.1 to this outcome. Those findings were considered as more critical than the partial positive bias of having additional courses marked by the students in comparison to those by the instructors. Such mismatches were interpreted as implying possible weaknesses or gaps in curriculum mapping, hence pointing to the need for improvement.

It is necessary to clarify here that, similar to the process for assessing the attainment level of the other outcomes, the A&E process for "SO4 – an ability to function on multidisciplinary teams" also included direct assessment by using rubrics in compulsory courses (i.e., in the senior-year compulsory Graduation Design Project course) as well as utilization of indirect assessment tools, namely the specific questions in several surveys targeting this specific SO (e.g. senior–exit, employers, internship–student, internship–employer surveys). In fact, assessment results from the last round of those surveys for level of attainment of SO4 at/or above satisfactory level were as follows (n: number of participants in the surveys): 58% according to employers' survey (n=7), 58% according to senior–exit survey (n=46), 100% according to internship–employers survey (n=125), and 90% according to internship–students survey (n=132). The students were also asked which multidisciplinary working groups/different professions they had a chance to work with during their compulsory summer internships, and they listed the following professions: civil, chemical, electrical, geomatics, and mechanical engineering as well as other engineering and professions (n=215). A group of senior-year students (n=16) was also asked if they worked with people from multi-disciplinary working groups/different professions while preparing their capstone engineering design project for the "CEV492/E Graduation Design Project" course; 75% of them confirmed that they had and added industrial, geological, and materials science and engineering to those listed above. Note that the surveys were not included in the curriculum maps presented here, as those maps were constructed to relate the contribution of the compulsory courses of the ITU EEUP's four-year curriculum to the attainment of the SOs.

In addition, the compulsory courses listed in Table 10.1 were those given by the instructors from the ITU Environmental Engineering Department. Courses offered in the program but given by the instructors from other departments were not included in the instructor-based map used in SO A&E. In their portfolio, however, the students related SO4 to "Computer-Aided Technical Drawing", "Materials Science", "Fluid Mechanics", and "Hydraulics", which were the compulsory courses of the program that were offered by the instructors from the Civil Engineering Department of ITU. In fact, the students referred specifically to problem-solving "in-class sessions" and "preparation of lab reports" for those courses during which "civil engineering and environmental engineering students worked in collaboration in mixed-teams", as stated by them in their sample portfolio.

In summary, the level of attainment of SO4 was assessed and evaluated by the program, but that is not fully reflected on the curriculum maps presented here. One final comment on SO4 is that in the full-length instructor-based map (data not presented here), several compulsory courses offered by the faculty members of the ITU EED were mapped to this outcome; however, these were not with the highest but with the intermediate level of contribution. Team work was encouraged or required in those courses; however, it was a challenge to accurately assess attainment of this outcome when considering the "multi-disciplinary" dimension of team work. Note that this challenging "multi-disciplinary" context has recently been removed by the ABET EAC itself from the expression of the SO on team work, and the revised student outcome for team-work is now described as "new SO5) an ability to function effectively on a team whose members together provide leadership, create a collaborative and inclusive environment, establish goals, plan tasks, and meet objectives" (ABET EAC, 2018; 2022).

The other outputs of comparing the student-based vs instructor-based maps might be listed as follows: The students related "Environmental Microbiology" to SO1, SO2, SO10; "Water Quality Management" to SO5, SO6, SO7; "Biological Processes" to SO1, SO3, SO5; and "Industrial Pollution Control" to SO5, SO7. On the other hand, relevant entries in the instructor-based map with the highest level of contribution and thus utilized in SO A&E were "Environmental Microbiology" for SO2; "Water Quality Management" for SO5; "Biological Processes" for SO1, SO5; and "Industrial Pollution Control" for SO5, SO7 (Table 10.1).

10.3.2 Examples from Students' Reasoning for "SO vs Course" Matching and Possible Inferences

Personal comments of the senior-year student volunteers in their sample team portfolio, were considered as a valuable, significant, and relevant source of information to facilitating the resolution of some additional possible reasons for the observed differences between the instructor- and student-based curriculum maps analyzed in this study (Table 10.1 and Figure 10.1). Out of several inputs from the students, those on two courses are cited here as examples. As the first example, the reasons described by the students for addressing the "CEV213/E Environmental Microbiology" course offered at the second-year as contributing to attainment of SO1, 2, 10 are as follows:

- Reason for matching the CEV213/E course with SO1 – apply knowledge of math, science and engineering: "This course was the first course in terms of applying math to environmental issues (i.e., bacterial growth-rate calculations and graphs, etc.). Also, viable cell counting methods (spread and pour plate techniques) were used to determine and calculate the number of microorganisms in the samples".
- Reason for matching the CEV213/E course with SO2 – experimentation and data handling: "In this course, students were doing experiments in the laboratory sessions carried out every week, preparing and submitting lab reports (10 detailed reports) on topics like using bright-field microscope, simple staining, Gram staining, investigation of filamentous microorganisms, etc. For example, in one of those weeks, the students brought surface-water samples they collected from their neighborhood and examined those in the lab to observe microorganisms. As a homework assignment; each student was expected to figure out what type of microorganisms were present in their samples, what were the differences from the findings of other

students, how the environmental conditions possibly affected the type of microorganisms".

- Reason for matching the CEV213/E course with SO10 – knowledge of contemporary issues: "Contemporary issues and their sustainable solutions were discussed in this course. For example, environmental pollution caused by petroleum leakage and its control in terrestrial and aquatic environments were covered. Also, some of the topics given to the students as homework assignments were related with contemporary issues; for example, fate and transport of micro-pollutants like antibiotics and hormones, and their impacts on aquatic microorganisms".

Considering the coverage and structure of this particular course with lab sessions where students conduct experiments on a weekly basis throughout the semester, this course was marked, as expected, both by the students and the instructors as contributing to attainment of SO2 – experimentation and data handling.

The students related this course to SO1 as well, while the instructors of this course did not relate it to that outcome in the context of assessment and evaluation: As mentioned above, only the compulsory courses with highest level of contribution (level-3) were considered in the SO A&E process, and this course was designated by the instructors as contributing to attainment of SO1 at an intermediate level (level-2). Note that this explanation is valid for several other mismatches seen in Table 10.1.

Moreover, the students related this particular course also to SO10, while this sophomore-year course was not mapped to that outcome by its instructors; content and scope of SO10 – knowledge of contemporary issues – were considered as requiring delivery and attainment of this outcome at an advanced year in the curriculum. Therefore, SO10 was related to senior-year courses in the instructor-based curriculum map as seen in Table 10.1.

Another case showing the details of the voluntary students' perspective in preparing their curriculum map is exemplified below through their reasoning for relating the third-year "CEV320/E Biological Processes" course as contributing to attainment of SO1, 3, 5:

- Reason for matching the CEV320/E course with SO1 – apply knowledge of math, science and engineering: students did not provide a specific comment on the relation of this course and SO1, but the generic entry they presented in the sample team portfolio for SO1 was as follows: "Almost all courses of the EEP serve for the achievement of ability to apply knowledge of mathematics, science and engineering. Physics, Chemistry and Mathematics courses constitute the basis for especially the courses offered at the 5-6-7-8th semesters providing students to function much more efficient in both design and processes lectures."
- Reason for matching the CEV320/E course with SO3 – design meeting needs within realistic constraints: "This course promoted the ability to compare wastewater treatment (WWT) alternatives and select the best option in accordance to the needs/demands of the public/society. Biological WWT system design was implemented and cost analysis was performed."

- Reason for matching the CEV320/E course with SO5 – identify, formulate, solve engineering problems: "With the basis from the 'Environmental Microbiology' course, microbial mechanisms were identified and classified. Students identified pollutants, categorized microorganisms for removal of pollutants from wastewater and from aquatic/terrestrial environments, acquired an ability to formulate related problems, learned and applied basic design of biological treatment units."

This final entry cited from the sample student (team) portfolio and reflecting the voluntary senior-year students' broader perspective across the curriculum is found to be of particular importance since it sets an example indicating the proper flow of the curriculum, pointing out the meaningful connection between the two courses offered in the 2nd and 3rd years of the Environmental Engineering undergraduate program, thus contributing to advancement of students' knowledge and skills as they progressed through the program, as intended.

10.3.3 Limitations and Need for Improvement

When the first research question of this study is revisited, the study is considered successful in providing a relevant and information-based answer. Analysis and interpretation of the results presented above through the initial "student portfolio and student curriculum mapping" exercise provide valuable inputs in structuring the following answer to that first research question: While preparing their academic portfolio, students' perspective about relating the courses to attainment of the outcomes partly differed from that of the instructors.

On the other hand, the presented study has its own limitations as listed below:

i. The sample student population size was limited to be representative of all senior-year students, which might have introduced a potential weakness of subjectivity and a self-selection bias in the portfolio and the curriculum map prepared by the voluntary students.

Revisiting the second research question of this study, the other limitations are as follows:

ii. The students did not specify the level of contribution (e.g., low, partly, high) when relating the courses to the SOs in their curriculum map, accordingly.

iii. The simplified curriculum maps presented in Table 10.1 include the compulsory courses with the highest level of contribution (level-3) but not those with lower or partial contributions (level-1 or level-2).

iv. This partly limits both the tool and the results in terms of showing the progress in achievement of the student outcomes throughout the four-year curriculum.

v. Moreover, no quantitative assessment data was present in the sample student (team) portfolio prepared by the voluntary senior-year students; therefore, it was not possible to use the portfolio directly in the SO A&E process.

Overall, those limitations implied the need to provide better guidance to the students while preparing their portfolios so that assessment of their outputs can be performed properly. In this context, recommendations for preparation of student portfolios and for their possible integration in the SO A&E process are given in the following section.

10.3.4 Projections for Preparation of Student Portfolios and Their Possible Use in the SO A&E Process

Different types of student portfolios and ways of preparing those have been presented in previous studies (Christy and Lima 1998). Student portfolios might be showcase, descriptive, evaluative, or composite type of compilations, each including different types and extents of documentation. Portfolios might be prepared in different ways: including all works of the students (non-selective) or only selected sample documents (selective). Another approach might be that, starting from the freshman year, students work on the same assignment and then further develop that assignment as they acquire more knowledge and skills in the junior and senior years. Students may then present their portfolios for review at the end of their senior year (Newcomer 1999). Alternatively, student portfolios covering the entire undergraduate education period are recommended as useful tools for presenting students' learning and reflections. Accordingly, incorporating such portfolios into curriculum assessment plans is suggested to contribute to the integrated outcome assessment schemes (Christy and Lima, 1998).

No matter what type of portfolio and how it is prepared, student portfolios are considered to encourage students to take greater responsibility for their own learning, engage in self-assessment of their own level of attainment of SOs, and take proactive roles in self-learning process (Christy and Lima, 1998). In addition to those benefits on individual basis, it is expected to be even more rewarding and valuable if portfolio preparation encompasses a whole cohort, especially the senior-year students. Accordingly, and also to address the first limitation of this current study listed above, a future implementation for improvement suggested here is to use the initial feedbacks from the this study and continue collecting student portfolios from a larger number – and if possible all – of the senior-year students of the Environmental Engineering Undergraduate Program.

In addition to the sample student works that are the primary tools and solid evidence used in direct assessment of SOs, student portfolios include valuable information reflecting students' perception and viewpoints. Accordingly, they might be integrated in the SOs A&E plan at the program level as a valuable source of information and/or as an additional assessment tool through which students' performance and advancements might be assessed and evaluated. However, for successful integration, it seems critical to address limitations such as those identified in this study, and provide better guidance to the students in preparing their academic portfolios. Mentoring and academic advising are also key factors that determine the satisfaction and self-improvement of the students (Ahmad, 2015). Considering that, and also to address some of the limitations determined in this study regarding the student portfolios, the student advisory courses that have been incorporated recently into the revised curricula of all engineering undergraduate programs at ITU, including the Environmental Engineering program, are considered to be effective platforms and good starting-points for providing better guidance to the students while preparing their academic portfolios. Those student advisory compulsory courses being offered to the ITU undergraduate students on a weekly basis on Wednesdays (between 15:30 and 17:30 for all registered students) are as follows (ITU Course Schedules – Undergraduate – Course Code: DAN, 2022):

- *DAN101/E Academic Advising*: freshman-year, fall and spring terms, 1-credit, 2-hours course incorporated into the ITU curricula as of 2017-18 Fall term, and then revised as the next listed course in 2021-22 Fall term,
- *DAN102/E Entrepreneurship and Career Advising*: freshman-year, spring term, non-credit, 2-hours course incorporated into the revised ITU curricula as of 2021-22 Fall term and beyond,
- *DAN301/E Career Advising*: junior-year, spring term, 1-credit, 2-hours course incorporated into the ITU curricula as of 2017-18 Fall term and being still in effect as of 2021-22 fall term and beyond.

In addition to those recently implemented student-advisory courses, each undergraduate student is also assigned to an academic advisor (full time faculty member) upon admittance to the ITU. Guidance and monitoring by the same advisor continues until graduation. Advisors follow academic data of their advisees and are able to monitor their progress through the online systems operated by the Office of Student Affairs and the Information Technologies Directorate at ITU. Accordingly, in order to successfully incorporate the use of student portfolios in assessing the attainment of the student outcomes, it is recommended here that the students are guided by their academic advisors for preparing their SPs. A core index might be suggested but not imposed to emphasize what is expected from them, and portfolio preparation will then be accomplished directly by the students themselves. They might be encouraged to demonstrate their achievements, considering not only the courses in the curriculum but also other activities (e.g., compulsory summer internships) that are supporting the attainment of the SOs.

Students' perceptions regarding the assessment process (information, procedures, criteria, conditions, etc.) are reported to be positively correlated with their learning process and thus assessment of the resulting outcomes (Gerritsen-van Leeuwenkamp et al., 2019). Therefore, knowledge-sharing regarding the tools, metrics, and processes being utilized in assessing the level of achievement of the SOs is expected to further improve the students' learning experiences and the SO assessment processes and also provide better guidance for the students while preparing their academic portfolios. Accordingly, in the two-hours/week student-advisory courses listed below and during the one-to-one appointments between the students and their assigned academic advisors, the students might be guided about the following topics:

- Why to prepare an academic (student) portfolio
- How portfolios can be benefitted
- How to submit student portfolios (e.g., paper vs. e-portfolios)

- The type of sample student works that might be included
- How those sample student works will be evaluated
- How the level of contribution of different courses to achievement of SOs might vary
- How to make use of analytical rubrics in assessing level of achievement of SOs, etc.

It is envisioned that students organize their own portfolios and perform all steps of their own work. They select relevant documents, collect items, and arrange data that might be assessed by their academic advisors and/or relevant parties in the department (e.g., accreditation coordination committee, curriculum development committee, SO A&E coordinators).

Moreover, assigned academic advisors and/or student advising courses' instructors might encourage the students in improving their self-learning competencies and support them in getting prepared for the professional life. Those faculty members, and particularly the academic advisors, might also provide formative feedback by reviewing the portfolios, meeting with the students to discuss their progress, and reporting their views. Formative feedback can be an integral component of teaching and learning processes to improve students' learning experiences and vice versa (Leenknecht et al., 2020).

When student portfolios prepared as described above are to be utilized in assessing the level of attainment of the student outcomes at program level, analytical rubrics that describe performance indicators (PI) and competency levels for each PI and SO should be used by the faculty members (student advisors, outcome assessment and evaluation coordinators, etc.). The quantitative assessment results to be obtained from student portfolios are expected to effectively contribute to assessing knowledge, competency, and skills acquired by the students (Soeiro et al., 2009). For programs that are planning to integrate student-advisory courses in their curriculum or continue providing academic advising to their undergraduate students, the results of this current study are encouraging in terms of starting or improving student advisory practices.

10.4   CONCLUDING REMARKS

The work was beneficial in terms of exploring the students' perspective of curriculum mapping – that is, students relating the compulsory courses of the Environmental Engineering undergraduate program to achievement of the student outcomes – and in terms of comparing the students' perspectives to the instructors' designations: While several aspects of the two maps by the voluntary senior-year students and the instructors of the courses were matching fully, there were some differences in the direction of students relating more courses to the SOs compared to the instructors, which reflects a broader student perception. Those additional student entries are considered to be indicating possible points of "students' self-learning", hence delivering the students' perspective in terms of displaying the strengths of the curriculum.

On the other hand, some other mismatches brought out by this study are considered to be pointing to some possible weaknesses/gaps in the curriculum-map utilized in SO assessments, hence referring to an opportunity for further improvement.

One technical limitation of this study was that the number of student volunteers in this initial exercise was quite low to represent a whole cohort. Yet it was decided to conduct the study with the available volunteers as the first case and initial effort to support further implementations. Therefore, for future, it is suggested to use the feedback from this study and continue collecting student portfolios but from a larger student population – from all senior-year students, if possible.

Finally, better guidance of the students is needed in portfolio preparation, which is recommended in this study to be accomplished through the student-advisory courses recently incorporated into the undergraduate curricula and also with the help of the academic advisors assigned to the undergraduate students upon admittance to the programs. The latter is expected to encourage the engineering students in their self-learning processes and in preparing for professional life.

ACKNOWLEDGEMENTS

We would like to acknowledge the work by Muratcan Başkurt – who was about to graduate at the time of data collection for the study and is currently a PhD student and a colleague – for the valuable contributions in preparing the voluntary senior-year students' sample portfolio and for help in compiling the students' entries for this study. Also, we extend our thanks to all the faculty members who took part in preparation of the instructor-based curriculum map. We would like to express our deepest condolences for our beloved friend and colleague Assoc. Prof. Didem Okutman Tas, whose exquisite work in this study is deeply appreciated. Without her precious contribution, this study would not have been possible. Her absence is a grave loss for our academia but her dedication to science and her students will continue to inspire us all. May she rest in peace.

REFERENCES

ABET. 2022. Official web-site, http://www.abet.org/; find an ABET-Accredited Program. Last accessed 12 Feb 2022. https://amspub.abet.org/aps/name-search?searchType=institution&keyword=Istanbul%20Technical%20University

ABET EAC, Baltimore, MD, USA. 2016. "Criteria for Accrediting Engineering Programs, 2017–2018". First accessed 27 January 2017, Last accessed 07 July 2020. https://www.abet.org/accreditation/accreditation-criteria/criteria-for-accrediting-engineering-programs-2017-2018/.

ABET EAC, Baltimore, MD, USA. 2018. "Criteria for Accrediting Engineering Programs, 2019–2020". Accessed 07 July 2020. https://www.abet.org/accreditation/accreditation-criteria/criteria-for-accrediting-engineering-programs-2019-2020/#GC3

ABET EAC, Baltimore, MD, USA. 2022. "Criteria for Accrediting Engineering Programs, 2022 – 2023". Last accessed 12 Feb 2022. https://www.abet.org/accreditation/accreditation-criteria/criteria-for-accrediting-engineering-programs-2022-2023/

ABET Webinar (2020). New outcomes: Transition as opportunity, by Daina Briedis, 9 April 2020 (18 min). First accessed 02 July 2020, Last accessed 12 Feb 2022. https://vimeo.com/405876096

Ahmad, S. Z. (2015). Evaluating student satisfaction of quality at international branch campuses. *Assessment & Evaluation in Higher Education*, *40*(4), 488–507. https://doi.org/10.1080/02602938.2014.925082

Aoudia, M., & Abu-Alqahsi, D. A. (2015). Curriculum redesign process for an industrial engineering program seeking ABET accreditation. *iJEP*. *5*(3), 45–52. http://dx.doi.org/10.3991/ijep.v5i3.4670

Bader, M., Burner, T., Iversen, S. H., & Varga, Z. (2019). Student perspectives on formative feedback as part of writing portfolios. *Assessment & Evaluation in Higher Education*, *44* (7), 1017–1028. doi: https://doi.org/10.1080/02602938.2018.1564811

Chan, Z., & Ho, S. (2019). Good and bad practices in rubrics: The perspectives of students and educators. *Assessment & Evaluation in Higher Education*, *44*(4), 533–545, https://doi.org/10.1080/02602938.2018.1522528

Christy, A. D., & Lima, M. (1998). The use of student portfolios in engineering instruction. *The Research Journal for Engineering Education*, *87*(2), 143–148. https://doi.org/10.1002/j.2168-9830.1998.tb00334.x

Dulekgurgen, E., Yangin-Gomec, C., Ozgun, O. K., Aydin, B., & Guven, H. (2018). Follow-up on assessment of student outcomes by senior-year design project and continuing to improve by performance indicator breakdown-based assessment. *International Journal of Engineering Pedagogy*, *8*(5), 19–28. https://doi.org/10.3991/ijep.v8i5.8148

Dulekgurgen, E., Taptik, I.Y, Aydın, A.F., & Hosni, M. (2019b). Power of continuous-knowledge sharing in preparing for successful accreditation reviews. *In BUA2019 Online Proceedings (Full Papers/Working Papers) - 5th Annual Conference of the Balkan Universities Association*, 16–18 April, 2019, Thessaloniki, Greece. http://bua2019.web.auth.gr/wp-content/uploads/2019/04/BUA2019_WorkingPapers.pdf; pp.146–152.

Dulekgurgen, E., Taptik, I.Y, Aydın, A.F., & Karaca, M. (2019a). International accreditation as a means of improving engineering education: the ITU - ABET experience. *In BUA2019 Online Proceedings (Full Papers/Working Papers) - 5th Annual Conference of the Balkan Universities Association*, 16–18 April, 2019, Thessaloniki, Greece. http://bua2019.web.auth.gr/wp-content/uploads/2019/04/BUA2019_WorkingPapers.pdf; pp.134–140.

ElAli, T. (2018). Innovation in engineering education: A proposed ABET course outcomes assessment portfolio. *International Journal of Innovative Research in Electronics and Communications (IJIREC)*, *5*(3), 28–34. http://dx.doi.org/10.20431/2349-4050.0503004

Felder, R. M., & Brent, R. (2003). Designing and teaching courses to satisfy the ABET Engineering Criteria. *Journal of Engineering Education*, *92*(1), 7–25. https://doi.org/10.1002/j.2168-9830.2003.tb00734.x

Gerritsen-van Leeuwenkamp, K. J., Joosten-ten Brinke, D., & Kester, L. (2019). Students' perceptions of assessment quality related to their learning approaches and learning outcomes. *Studies in Educational Evaluation*, *63*, 72–82. https://doi.org/10.1016/j.stueduc.2019.07.005

Gredler, M. E. (1995). Implications of portfolio assessment for program evaluation. *Studies in Educational Evaluation*, *21*, 432–437. https://doi.org/10.1016/0191-491X(95)00024-O

ITU Course Schedules – Undergraduate – Course Code: DAN. (2022). "Course Schedules (itu.edu.tr)", Last accessed 12 Feb 2022. https://www.sis.itu.edu.tr/EN/student/course-schedules/course-schedules.php?seviye=LS&derskodu=DAN

ITU EED. (2022). Istanbul Technical University, Environmental Engineering Department official accreditation info web page: "ABET Accreditation (itu.edu.tr)" (2022). Last accessed 12 Feb 2022. https://cevre.itu.edu.tr/en/education/accreditation/abet-accreditation

Kumpas-Lenk, K., Eisenschmidt, E., & Veispak, A. (2018). Does the design of learning outcomes matter from students' perspective? *Studies in Educational Evaluation*, *59*, 179–186. https://doi.org/10.1016/j.stueduc.2018.07.008

Lam, R. (2014). Promoting self-regulated learning through portfolio assessment: Testimony and recommendations. *Assessment & Evaluation in Higher Education*, 39(6), 699–714. https://doi.org/10.1080/02602938.2013.862211

Leenknecht, M., Wijnia, L., Köhlen, M., Fryer, L., Rikers, R., & Loyens, S. (2020). Formative assessment as practice: The role of students' motivation. *Assessment & Evaluation in Higher Education, (Ahead-of-Print)*, 1–20. https://doi.org/10.1080/02602938.2020.1765228

Mintz, K., & Tal, T. (2014). Sustainability in higher education courses: Multiple learning outcomes. *Studies in Educational Evaluation*, *41*, 113–123. https://doi.org/10.1016/j.stueduc.2013.11.003

Morgan, S. M., Cross, W. B., & Rossow, M. P. (2000). Innovative outcome portfolios for ABET Assessment. Proceeding of the 2000 Annual Conference, St. Louis, Missouri, USA, pp. 5.361.1–5.361.7.

Newcomer, J. L. (1999). The development of a multi-course project for engineering technology students. *FIE'99 Frontiers in Education. 29th Annual Frontiers in Education Conference. Designing the Future of Science and Engineering Education. Conference Proceedings (IEEE Cat. No.99CH37011*, San Juan, Puerto Rico, USA, 1999, pp. 13A9/5 vol.3-, doi: 10.1109/FIE.1999.840336

Scales, K., Owen, C., Shiohare, S., & Leonard, M. (1998). Preparing for program accreditation review under ABET Engineering Criteria 2000: Choosing outcome indicators. *The Research Journal for Engineering Education*, *87*(3), 207–210. https://doi.org/10.1002/j.2168-9830.1998.tb00342.x

Shafi, A., Saeed, S., Bamarouf, Y. A., Iqbal, S. Z., Min-Allah, N., & Alqahtani, M. A. (2019). Student outcomes assessment methodology for ABET Accreditation: A case study of computer science and computer information systems programs. *IEEE-ACCESS 7*: 13653–13667. https://doi.org/10.1109/ACCESS.2019.2894066

Sin, C., Tavares, O., & Amaral, A. (2017). The impact of programme accreditation on Portuguese higher education provision. *Assessment & Evaluation in Higher Education*, *42*(6), 860–871. https://doi.org/10.1080/02602938.2016.1203860

Soeiro, A., Martins, I., & Correia, T. (2009). E-portfolios and assessment: A case study for civil engineering. Proceeding of the International Conference of Education, Research and Innovation (ICERI 2009), Madrid, Spain, 1–5.

Trevisan, M. S., Davis, D. C., Calkins, D. E., & Gentili, K. L. (1999). Designing sound scoring criteria for assessing student performance. *Journal of Engineering Education*, *88*(1), 79–85. https://doi.org/10.1002/j.2168-9830.1999.tb00415.x

UDel CTAL. (2020). University of Delaware – Center for Teaching and Assessment of Learning official web-site, "Curriculum Mapping" 2020. Accessed 07 July 2020. https://ctal.udel.edu/resources-2/mapping/

11 Role of Learning Management Systems in Online Learning
A Case Study

N. Subadra, B. Nagamani, Aryan Kodte,
and Ananya Chepuri
Geethanjali College of Engineering and Technology
Hyderabad, Telangana, India

CONTENTS

11.1   INTRODUCTION

The world is facing the grave peril of the COVID-19 pandemic. It walloped multiple countries to the extent that governments imposed lockdowns and restrictions in order to curb the spread of the deadly virus. It led to a decline in economies, and many countries suffered a huge shortfall in their revenues as taxpayers could not pay the tax. Countries like the USA, India, Brazil, European nations, and others silently watched the cataclysm that shook the supply system causing it to reach impasse. COVID-19 made clear that consequences of exploiting nature are inevitable, and it has reminded us of the adage: "Health is not valued till sickness comes". The world has caught a glimpse of present-day circumstances and countless revolutionary changes in the 21st century. Drastic reduction in pollution allowed nature to rejuvenate and refuel its natural cycle. The health sector is being prioritized by governments. Technology is advancing at a rapid pace, and shifts to online mode in both academic and non-academic purposes have taken a quantum leap. As COVID-19 started escalating and creating chaos, the impact of the pandemic on education was serious and forced an change in the curriculum, disturbing the entire education industry. It also affected the teaching progress, which is likely to have a far-reaching impact for the future of students around the world. Learning and

DOI: 10.1201/9781003083160-15

teaching environments have been forced to change overnight, with dramatic adjustments required by all stakeholders (students, teachers, administrators, parents, etc.). It has created a sense of unpredictability and apprehension about what's going to happen by increasing stress and having ill effects on the psychological well-being of students.

Technology has played an indispensable role in everyone's life. The world is heading towards Fifth Generation computer technology where data is as valuable as gold. It has become a constituent element in today's generation, and humans have been making use of the technological networks for the sake of necessity. The richness of technology-based education is a blessing, which is a supporting factor in improving students' achievements. Online learning has become indispensable during the COVID-19 pandemic when social contact is minimal.

Online learning can expand learning beyond classrooms. It provides time and scope for the learners to reflect, understand, and adapt to changing needs of learning as much as they want. This facility in online learning provides more time for discussion, both online and face to face. It provides more time for performing skill-based activities. Online learning eliminates commuting problems, saves time, and reduces the cost incurred in learning opportunities in comparison to classroom or face-to-face learning.

Eventually people learn by doing tasks – through practice. Learning by doing is the present norm. Much of what is needed on a job is learned through observation, sharing, and support from the peers. Effective online learning focuses on providing adequate models for practice and feedback on real-time achievements. A well-developed online learning strategy provides innumerable opportunities for social interaction and collaboration so that the learners are connected with other learners across the globe without geographical restrictions.

They not only connect with other learners, they can also connect with other tutors, experts, and specialists and can share relevant material both in text and any media formats. Choice, convenience, and computability are the success factors in online learning. Myriad technologies have made online learning achievable, but every so often they have challenges that a student meets with disengagement and boredom.

Lectures delivered by professors often make students lose interest and track of time, which is not practical. Students find it tedious and laborious to visualize the contents, and that often breaks the stream of continuity.

Students find it strenuous in front of the screen, and it severely impacts cognitive facilitation of promoting student learning. Students face a huge issue of clearing up doubtful understanding, and this itself defeats the very purpose of two-way interaction.

Students sabotage their own achievements as they cannot reach full potential in proficiency due to lack of group-based activities. Online classes can never supersede traditional sessions, and there is a stumbling block that cannot be bridged digitally. Also, students have technical issues in logging in to the online classrooms with proper internet access, thereby creating a digital divide.

Moreover, there can be audio and video problems and errors in downloading files, which take time to troubleshoot and debug at the same time, becoming barriers for online learning. Inadequate personalization of the learning processes can create an imbalance in meeting educational requirements.

It is a well-known fact that the students' critical thinking skills increase during the learning process. Education modules are drawn up in a way that creates an environment for the students to set foot into this journey of growth and development.

Young, energetic students brimming with positivity, hope, and ambition and fueled with passion and willingness to work harder are waiting at one end, and at the other end is their bright future, opportunities, and career. It is not possible for online technology to build a bridge between these two ends. Moreover, analogy and reasoning skills require practicality, and this is attainable and plausible only when a student steps out into the real and physical world. The ability of students to manage their affairs and to think about achieving their learning goals is not everyone's strength.

Learners can be broadly classified into auditory, visual, and kinesthetic. Auditory learners are comfortable listening to audio tapes and prefer listening to reading. A visual learner prefers to learn through visuals and or text. Kinesthetic learners enjoy learning through visiting various websites, incorporating online research, and texting to submit the course material. Most learning management systems provide several opportunities to accommodate diverse learners. Azhar and Iqbal (2018) discussed the growing significance of the role of technology in education and the accessibility of various open source material for the teachers.

Lombardi (2019) opined that a better shared approach in learning will result in larger, deeper and better understanding in creating and sharing knowledge.

Technological advancement has changed businesses, markets, innovations, and education. The advent of online teaching has provided opportunities to teachers to research freely available and easy-to-use digital tools, open-source course management, an/or learning management systems.

Kim and Bonk (2006), in their survey on trends in online education, concluded that as the demand for online learning is increasing, instructors should know how to provide, moderate, and facilitate learning and how to develop a collaborative learning atmosphere that reduces the digital divide between the teacher and the learners and among learners.

Oblinger and Oblinger (2005), emphasizes that educational technology plays a noteworthy role in bringing changes in education. To cope with this significantly advancing environment, deep-seated strategies are essential to fulfill the needs of higher education at present and in the future. Information technology has brought in several changes in education and in other sectors. It creates new enterprises.

These newly created enterprises serve as models that help higher education to serve new and diverse groups of students, in large numbers, to provide better learning outcomes. The use of these technologies helps in moving towards learner-centric and learner-generated content. The present paper focuses on the advantages of two learning management systems, Edmodo and Google Classroom.

A learning management system (LMS) is a software application for the administration, documentation, tracking, reporting, automation, and delivery of educational courses, training programs, or learning and development programs. The learning management system concept emerged directly from e-learning.

The present research was conducted during online teaching in Geethanjali College of Engineering and Technology, Hyderabad for teaching English and Mathematics in B.Tech. first year II semester.

Edmodo is a less-known learning management system developed by Nick Borg, Jeff O'Hara, and Crystal Hutter in 2008. This free and secure LMS is a mirror reflection of an offline classroom that accommodates all the classroom needs. The various features available make learners comfortable and at home. This LMS can be used effectively by teachers and students. Parents – important stakeholders – also understand the progress of their wards at regular intervals. This is secure and safe because the teachers create the Edmodo account, and the unique code obtained will be shared with the students. One teacher can create several classrooms and every classroom has a unique code for the students to join.

The various features available on Edmodo make teaching comfortable. The calendar page helps students know the topics to be dealt in the next class. This prepares the students mentally and provides an opportunity for advance reading. The quizzes and assignment options available on the platform help the learners to gauge the understanding and allows the teachers to know the cognitive level of the students.

Edmodo provides opportunities to create subgroups, such as Reading Club, Writing Club, etc., for encouraging the students to read, write, and share their learning. This creates an encouraging learning circle. Students can post book reviews and discuss the author's style and characterization. This will inculcate, promote, and encourage reading and writing habits among the students. They also develop the ability to interpret, analyze, and look at a particular aspect through various perspectives.

Grading has an important role to play in teaching–learning process. In engineering colleges, according to the university's rules, assignments are given to students at the end of every unit to self-test their understanding of the important concepts dealt in the unit. These assignments also carry marks and are added into the internal assessment, which is 25% of the main grade obtained in the end-of-semester exam. Submitting assignments on time is also considered important, as late submission results in a reduction in marks.

A notable feature of Edmodo is that it does not accept any assignment sent after the deadline set by the teacher. Whenever an assignment is posted by the teacher, it is displayed on the account through which the teacher has created the Edmodo platform. This will help the teacher identify the learners who have submitted the assignment. This feature develops sense of responsibility among the students, and they tend to submit tasks on time.

Parents and teachers can establish a closed networked relationship among themselves. Parents can track their ward's progress by logging on to the LMS through a separate code, which is generated solely for them.

The inbuilt discussion polls provide the learners with interesting opportunities to participate in online discussions. The discussion thread can be continued, and every student can add their opinion on a given topic. Students can get clarifications through their peers and need not always depend on the teacher to answer a few questions.

The teacher has the opportunity to follow the thread of discussion and clarification and can step in at the right time to put the discussion on track. Suggestions, comments, or remarks posted by the teacher are open. This feature of open feedback helps all the other learners.

Students are encouraged by receiving badges recognizing their performance through Edmodo's "my student badges" option, which is available under "progress" on the home page. Teachers also receive badges for completing their profiles for being the first person to use Edmodo either from their school or college or university and or workplace.

Google Classroom is a free web service developed by Google and released on August 12, 2014. Google Classroom can be integrated with Google Meet. The teacher invites the students to join the classroom through a unique code generated. This feature is similar to that of Edmodo. One classroom accommodates several classes by one teacher.

However, subgroups to promote discussion on students' reading and writing habits cannot be created. The calendar feature gives prior information to students on the topic that will be discussed by the teacher. This facility prepares the learners to be ready for the class and provides opportunity for the learner to know the basic information on the topic.

The teacher is notified when assignments are submitted by the students. If a student submits the assignment after the due date set by the teacher, the assignment is accepted but labelled as a late submission. This is a ready reference for the teacher to know the details of the students who have not submitted the assignments in time. The rubric for evaluating the assignments can be given to the students. Hence students can know the areas to improve and strengthen.

Teacher comments and marks are open to all the learners of that particular class. Any four assignments can be checked for the percentage of plagiarism in the assignment. The teacher can openly post their comments on plagiarism in the assignments, which will caution the other students. Thus, the habit of copy/paste is curtailed at an early stage.

The students are trained in answering objective questions through the quizzes feature. Quizzes can be either created or teachers can use already integrated Google forms or docs or drawings. Additional topical information can be posted. The question option can be used as a discussion platform to increase peer learning and collaboration.

Both Google Classroom and Edmodo are available on Android and iOs. As part of G Suite for Education package, Google Classroom is incorporated with other Google applications: Gmail, Google Drive, Google Docs, Sheets and all other applications of Google such as Slides, Calendar, and YouTube.

Adi (2019), explained that the learning activities taken up through these LMS are convenient for the learners as these are similar to other social networking sites with which they are familiar with. Nagamani (2020) summarized the various opportunities and challenges in using LMS.

Pappas (2015), in his paper, opined that Google Classroom provides great support to the students in creating opportunities for communication and collaboration with other and gives immediate feedback from the teacher.

Shaharanee et al. (2016) analyzed the effectiveness of Google Classroom for data mining. Adas and Bakir (2013), in their paper, emphasized the role of ICT tools in improving writing abilities among students.

Feri and Erlinda (2014) highlighted that learner autonomy and collaborative learning can be enhanced through ICT tools. Nagamani et al., (2021) explained various reasons for the increasing use of technology and technology-based tools in online learning.

Target Audience: I. B. Tech I semester and II semester students of Electronics and Communications Engineering (ECE), Mechanical Engineering (ME), and Computer Science Engineering (CSE) students of Geethanjali College of Engineering and Technology, Hyderabad.

Method: Oral interviews with the users of Edmodo and Google Classroom were taken. A questionnaire comprising of closed-ended and open-ended questions was also circulated to the students. Nearly 60 students from each branch of engineering were considered for this research process. The oral interviews focused on the feasibility of the learning management systems (both Edmodo and Google Classroom) from the learners' perspective, features, the role of these LMSs in increasing their understanding of the subject, and their satisfaction levels in using the learning management systems.

The questionnaire, comprising the following questions, mainly focused on features and satisfaction level of learners on both LMSs.

- How does Edmodo help the students?
- Which feature of Edmodo do you like the most?
- Does Edmodo provide peer learning opportunities?
- Do you get notifications when something is posted on Edmodo?
- How satisfied are you by using Edmodo?
- How does Google Classroom help the students?
- Which feature of Google Classroom do you like the most?
- Does Google Classroom provide peer learning opportunities?
- Do you get notifications when something is posted on Google Classroom?
- How satisfied are you by using Google Classroom?

11.2 ANALYSIS

Based on the responses provided by the students as part of oral interviews, the majority of students agreed that both Edmodo and Google Classroom were very useful, more so in remote, or online, learning. Since the learners are digital natives, they were very comfortable with the user-friendly LMSs. While answering questions on the available features, most of the students appreciated the discussion board or create-a-question option in Google Classroom.

They expressed that it created a lot of discussion on a given topic. Some students said that shy students gained more confidence in uploading the assignments on the Edmodo or on Google Classroom than when giving the same to the teachers face to face. The majority of learners responded that since the LMSs are easy to use and do not require greater internet speed, they were satisfied. Some learners expressed that they did not want to see their names in the not submitted list or late submission and cultivated the habit of posting the assignments on time.

See Figures 11.1 to 11.4 for results of the surveys.

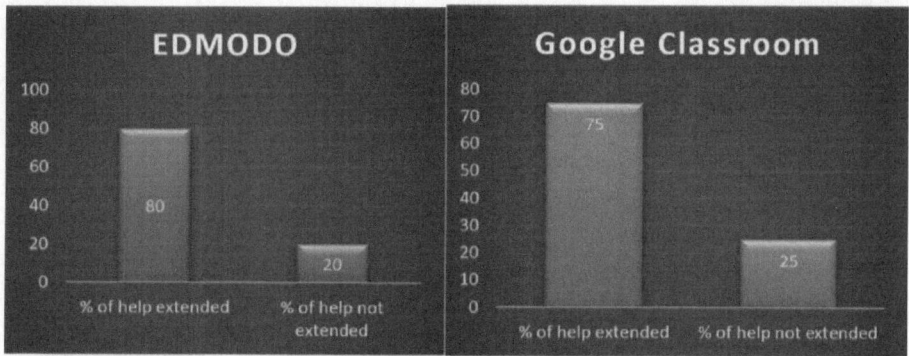

FIGURE 11.1 Percentage of support extended by both LMSs: 80% of students expressed that Edmodo has provided lot of support while learning online. However, 75% of students opined that Google classroom has provided lot of support while learning online.

All the students said that they receive notifications on any new post in both LMSs. Edmodo has no option to block these notifications. However, Google Classroom has the feature to block notifications.

11.3 RESULTS AND DISCUSSIONS

After a thorough analysis of the responses, both LMSs were found to be extremely useful in supplementing the learning material and peer learning, and the users are satisfied with the features. Nevertheless, learners expressed that it is difficult to read if a picture of text is uploaded on Edmodo. The question of legibility arises. Thus, the students' only option was to upload a Word document. This is faced by students who used mobile phones.

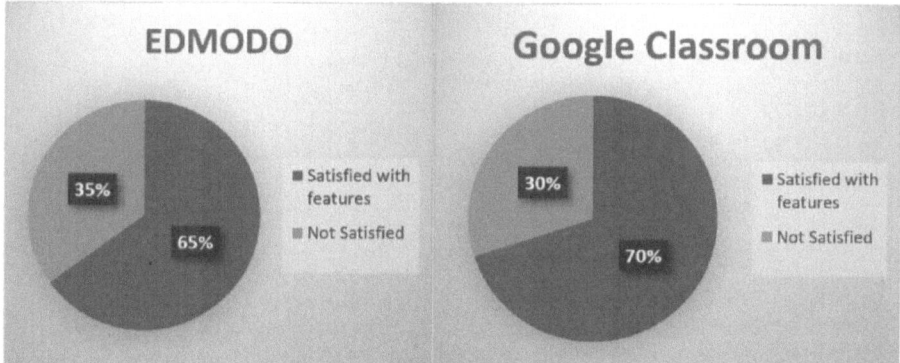

FIGURE 11.2 Percentage of satisfied students with features by both LMSs: 65% of the students were satisfied by the features of Edmodo, whereas, 70% of the students were satisfied by the features of Google classroom.

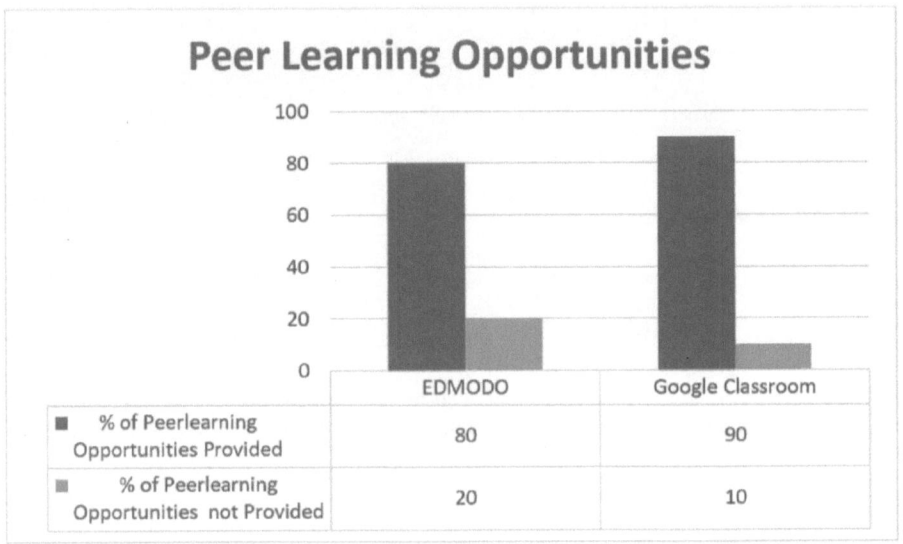

FIGURE 11.3 Percentage of peer learning opportunities provided by both LMSs. The peer leaning opportunities provided by Edmodo satisfied 80% of students and Google classroom satisfied 90%.

Google Classroom does not have an inbuilt facility to mail the parents regarding their wards' performance. It does not permit access from multiple domains.

A lot of issues are attached to online learning, but one cannot deny the fact that it helped immensely in imparting education in times of unprecedented crisis. We can always find solutions to fix these difficulties. Educators must come up with a plan of action to make a student-centered environment in order to have communication and group-based activities.

Educators can devote 10 to 15 minutes at the end of each class by giving personal attention to students so that they can easily adapt to this form of learning and pick

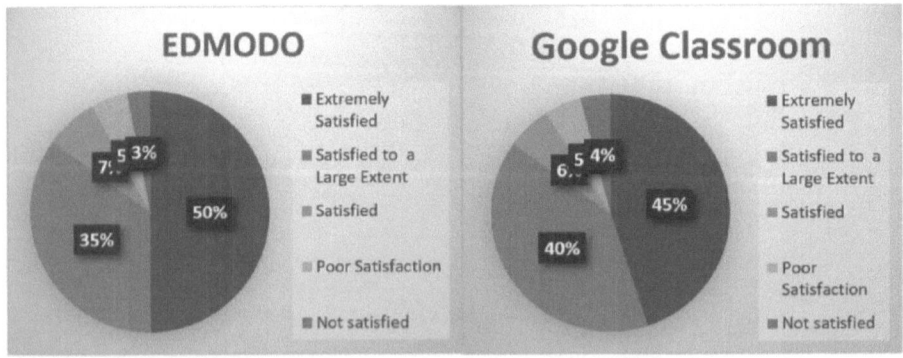

FIGURE 11.4 Satisfaction percentage of the students provided by both LMSs: 85% of students were satisfied extremely and to a large extent by both LMSs.

up the threads of enthusiasm and astuteness. Motivating the students on a timely basis will bring the best out of them. Effective changes in the course objectives can enhance learning productivity.

Receiving constructive feedback from the students will assist the educators in knowing some of the challenges the students face. Some teaching strategies (lectures, case studies, debates, discussions, experiential learning, brainstorming sessions, games, drills, etc.) can be used online to facilitate effective and efficient teaching and learning practices. Finally, identifying a suitable LMS will help the educators in achieving the outcomes.

11.4 CONCLUSION

It is predicted that by the end of this decade more than half of the world will be digitally connected. Technology is changing the social and societal norms and structures. The higher education sector is the first and is being gravely interrupted due to technological changes and innovations. Traditional on-site or face-to-face teaching is replaced or reduced by teaching and learning across digital spaces. Since digital technologies are exploding, it is important that both teachers and learners explore and embrace these changes to meet the needs of the students.

The paradigm shift from face-to-face learning to online learning or blended learning has given new roles and responsibilities to the teachers, learners, and parents. Learners have to take up greater responsibility and interest and participate enthusiastically in the learning process. This will contribute and promote a culture of learning. Academicians have to adopt new technological innovations and adapt to the changing scenarios. Both educators and practitioners of engineering should be accustomed to diverse environments. This will help them to plan and manage any change in the learning environment. Parents as stakeholders should assume greater responsibility and ensure that their children are prepared for life after education, be it employment or higher studies.

Based on the results of research, it is clear that online technology played significant part in these difficult times. It shifted the role of the teacher from being a knowledge provider to a coach and mentor. Online classes are a necessity now, but the consequences cannot be ignored. Given these challenges, education can be driven towards a learner-centered approach. Students also need to have a certain amount of digital literacy to cope with hardship.

The COVID-19 pandemic is proving to be creating chaos, giving an opportunity for reshaping the teaching and learning environments, and bringing much-needed innovation and change. The utility of both the LMSs is nearly the same. The teacher can use their discretion based on the subject and the students' learning needs. Both these LMSs provide ample opportunities to create a favorable platform to learn, participate, and perform.

REFERENCES

Adas, D., & Bakir, A. (2013). Writing difficulties and new solutions: Blended learning as an approach to improve writing abilities. International Journal of Humanities and Social Science, 3(9), 254–266.

Adi, S. S. (2019). Utilizing Edmodo and Google Classroom for facilitating blended learning, Advances in Social Science. Education and Humanities Research, 387, 159–163.

Azhar, K. A., and Iqbal, N. (2018). Effectiveness of Google classroom: Teachers' perceptions, Prizren Social Science Journal, 2(2), 52–66.

Lombardi, P. (2019). Instructional Methods, Strategies, and Technologies to Meet the Needs of All Learners. Retrieved from https://granite.pressbooks.pub/teachingdiverselearners/

Feri, Z. O., & Erlinda, R. (2014). Building students' learning autonomy through collaborative learning to develop their language awareness. Proceedings the 5th International Seminar on English Language Teaching, 2, 518–523.

Kim, K., & Bonk, C. (2006). The future of online teaching and learning in higher education: The survey says. Educause Quarterly, no. 4. Retrieved from net.educause.edu/ir/library/pdf/eqm0644.pdf

Nagamani, B. (2020). Flip-Flop with ICT. The English CLASSROOM, Bi-Annual Journal published by Regional Institute of English, South India, ISSN No: 2250-2831, vol. 22, June & December 2020.

Nagamani, B., Subhadra, N., & Kodte, A. (2021). Implication of technology: Past and during COVID-19. Journal of Huazhong University of Science and Technology, 50(7), 1–10.

Oblinger, D. G., & Oblinger, J. L. (2005), Educating the Net Generation. Educause.

Pappas, C. (2015). Google Classroom review: Pros and cons of using Google Classroom in e-learning. Retrieved from https://elearningindustry.com/google-classroom-review-pros-and-cons-of-using-google-classroomin-elearning.

Shaharanee, N. M., Jamil, J. M., and Rodzi, S. S. M. (2016). Google Classroom as a tool for active learning. *AIP Conference Proceedings*, 1761(1), 20069.

Index